Change Detection and Input Design in Dynamical Systems

Feza Kerestecioğlu
Control Systems Centre, UMIST, UK
Now at: *Boğaziçi University, Turkey*

RESEARCH STUDIES PRESS LTD.
Taunton, Somerset, England

JOHN WILEY & SONS INC.
New York · Chichester · Toronto · Brisbane · Singapore

RESEARCH STUDIES PRESS LTD.
24 Belvedere Road, Taunton, Somerset, England TA1 1HD

Copyright © 1993 by Research Studies Press Ltd.

All rights reserved.

No part of this book may be reproduced by any means,
nor transmitted, nor translated into a machine language
without the written permission of the publisher.

Marketing and Distribution:

Australia and New Zealand:
Jacaranda Wiley Ltd.
GPO Box 859, Brisbane, Queensland 4001, Australia

Canada:
JOHN WILEY & SONS CANADA LIMITED
22 Worcester Road, Rexdale, Ontario, Canada

Europe, Africa, Middle East and Japan:
JOHN WILEY & SONS LIMITED
Baffins Lane, Chichester, West Sussex, England

North and South America:
JOHN WILEY & SONS INC.
605 Third Avenue, New York, NY 10158, USA

South East Asia:
JOHN WILEY & SONS (SEA) PTE LTD.
37 Jalan Pemimpin 05-04
Block B Union Industrial Building, Singapore 2057

Library of Congress Cataloging-in-Publication Data

Kerestecioğlu, Feza, 1961–
    Change detection and input design in dynamical systems / Feza
Kerestecioğlu.
      p. cm. — (UMIST Control Systems Centre series ; 1)
    Includes bibliographical references and index.
    ISBN 0-86380-152-8 — ISBN 0-471-94173-5 (Wiley)
    1. Adaptive control systems. 2. Adaptive signal processing.
I. Title. II. Series:
TJ217.K47 1993
629.8'36—dc20   93-2374
                                                  CIP

British Library Cataloguing in Publication Data

A catalogue record for this book
is available from the British Library.

ISBN 0 86380 152 8 (Research Studies Press Ltd.)
ISBN 0 471 94173 5 (John Wiley & Sons Inc.)

Printed in Great Britain by SRP Ltd., Exeter

Editorial Foreword

For over 25 years the Control Systems Centre at UMIST has been at the forefront of research in the key areas of control, signal processing and - more recently - information technology. It is the intention of this book series to make more widely and more speedily available the best of the current contributions in these areas emerging from the Centre and, where appropriate, from collaborating individuals or organisations. Each volume will aim to present new results and an up-to-date survey of a particular area of current or potential importance to researchers and the wider engineering community.

Martin Zarrop
Peter Wellstead

Control Systems Centre
Department of Electrical
Engineering and Electronics
University of Manchester
Institute of Science and Technology

Preface

This work is concerned with the detection and diagnosis of abrupt changes in the dynamics of single input single output stochastic systems and the design of auxiliary inputs for this purpose.

First, the problem is put in a decision theoretic framework and the concepts of classical decision theory (where no input is available to the decision maker) are augmented to accommodate the input design as well as the decision rules for change detection. It is shown that some important properties of the decision rules related to the classical case still hold.

The sequential probability ratio test (SPRT) which arises as a suboptimal test within this framework is investigated in detail. The Fundamental Identity of sequential analysis and the performance measures of the test are derived for the case of autoregressive models. The application of SPRT to the change detection problem is discussed and the properties of an associated cumulative sum (CUSUM) test are analyzed.

In designing inputs to improve the performance of the CUSUM test the design objectives are taken as not only to minimize the average detection delay but also to ensure a specified false alarm rate. Both offline and online generated signals are considered.

The detection mechanism as well as the input design techniques are extended to the multihypothesis case.

Acknowledgements

I wish to express my deep gratitude to Dr. Martin B. Zarrop for his insightful guidance and many enthusiastic discussions. His understanding and optimism have always been a great source of comfort to me.

I have very much benefited from the discussions with other members of the Self-Tuning Group in Control Systems Centre, among whom I especially would like to mention Dr. X.J. Zhang, A. Jenssen, Dr. P. Wellstead and J.J. Troyas. I thank them all.

I also would like to acknowledge the sponsorship of Boğaziçi University, Istanbul, for my studies in UMIST.

Finally, I do not think I can thank enough my dear wife Mine for her everlasting support and for her sharing both happy and sad moments throughout the course of this work with me.

Contents

Notation

1 Introduction   **1**
    1.1 Change Detection 1
       1.1.1 Generation of residuals 2
       1.1.2 Decision making 4
    1.2 Input Design 7
    1.3 Review and Original Contributions 9

2 A Decision Theoretic Approach to Change Detection   **13**
    2.1 Introduction 14
    2.2 Decision Rules Including Input Law 15
       2.2.1 Elements of sequential decision problem 15
       2.2.2 Decision rules 17
       2.2.3 Bayes risk 20
    2.3 Dynamic Programming Solution for the Decision Rules 20
       2.3.1 Optimal terminal decision rule 20
       2.3.2 Augmented stopping rule: Truncated case 23
       2.3.3 The untruncated case 25
    2.4 A Special Case 27
    2.5 An Example 33
    2.6 Conclusions 39

3 The Sequential Probability Ratio Test — 40
- 3.1 Definition — 40
- 3.2 Performance and Properties — 43
  - 3.2.1 Fundamental identity — 43
  - 3.2.2 ASN and OC functions — 44
- 3.3 Examples — 46
- 3.4 Conclusions — 49

4 Sequential Analysis of Autoregressive Processes — 51
- 4.1 Introduction — 52
- 4.2 Some Lemmas — 54
  - 4.2.1 Asymptotic behaviour of conditional log likelihood ratio — 54
  - 4.2.2 Properties of $\lambda(t)$ — 62
- 4.3 An Analogue of Fundamental Identity — 66
- 4.4 ASN and OC Function — 67
- 4.5 The Exact Log Likelihood Ratio — 69
- 4.6 Simulation Examples — 72
- 4.7 Conclusions — 75
- Appendix 4.A Derivation of Equation (4.42) — 77
- Appendix 4.B Derivation of Relation (4.58) — 79

5 Change Detection in Dynamical Systems — 80
- 5.1 CUSUM Test — 80
  - 5.1.1 Rationale and definition — 80
  - 5.1.2 Application to CARMA models — 83
- 5.2 Performance Measures of CUSUM Test — 85
- 5.3 Examples — 88
- 5.4 Conclusions — 91

6 Input Design for Change Detection   92

    6.1   Introduction 93

    6.2   Problem Definition 94

    6.3   Offline Inputs 96

         6.3.1   Problem refinement 96

         6.3.2   Power constrained inputs 98

    6.4   Online Inputs 105

         6.4.1   Problem refinement 105

         6.4.2   A suboptimal solution 108

    6.5   Simulation Examples 112

    6.6   Conclusions 116

    Appendix 6.A Proofs of Lemmas 6.1 and 6.2 116

7 Multihypothesis Change Detection   121

    7.1   Multihypothesis SPRT 121

    7.2   Multihypothesis CUSUM Test 124

    7.3   Multihypothesis Input Design 130

         7.3.1   Problem definition 130

         7.3.2   Offline inputs 133

         7.3.3   Online inputs 138

    7.4   Conclusions 140

References   141

Index   149

List of Figures

1.1 Basic stages in a change detection scheme 3

2.1 The function $W_k(p, \varphi)$ and decision thresholds 32

4.1 ASN of SPRT in testing first order AR parameter 74
4.2 OC function of SPRT in testing first order AR parameter 75

5.1 Typical behaviour of SPRT and CUSUM test in detecting a change . 81
5.2 ARL of the CUSUM test in testing first order AR parameter 89

6.1 The functions $|T_0(e^{j\omega})|^2$ and $|T_1(e^{j\omega})|^2$ in Example 6.1 103

7.1 Typical behaviour of the cumulative sums $\mathcal{S}_k(i,j)$ and $\bar{\mathcal{S}}_k(i,j)$ in Example 7.1 128

List of Tables

2.1 Decision variables in Example 2.2 if $y(i) = 1$ for all $i \leq k$ 38

3.1 Performance of SPRT in testing the mean of a normal distribution . . 47

4.1 Performance of SPRT in testing second order AR parameters 76

5.1 Performance of CUSUM test in testing ARMA parameters 91

6.1 Estimated average detection delays under offline inputs 113
6.2 Estimated mean times between false alarms under offline inputs . . . 113
6.3 Estimated detection and false alarm performance under online inputs 115

7.1 Estimated performances of Zhang's procedure and multihypothesis CUSUM test in testing the mean of a Gaussian random variable . . . 127
7.2 Estimated detection and false alarm performances of the multihypothesis CUSUM test under offline inputs 138

Notation

Lowercase and uppercase bold symbols are used for vectors and matrices, respectively. All nonbold characters refer to scalar quantities or sequences or sets of them. A prime ($'$) denotes derivative; the transpose of an array is denoted by the superscript (^{T}).

The end of a proof is marked with □.

List of Symbols

$A(\cdot)$ ($A_i(\cdot)$)	polynomial defining the AR part of the model (under $\mathcal{H}_i$)
$B(\cdot)$ ($B_i(\cdot)$)	polynomial defining the input dynamics (under $\mathcal{H}_i$)
$c(\cdot, \cdot, \cdot)$	cost of sampling and inputs
$C(\cdot)$ ($C_i(\cdot)$)	polynomial defining the MA part of the model (under $\mathcal{H}_i$)
d	terminal decision function
D	terminal decision rule
e_i	one-step-ahead prediction error based on $\mathcal{H}_i$
$E\{\cdot\}$	expectation operator
$f(\cdot)$ ($f_i(\cdot)$)	probability density function of y (under $\mathcal{H}_i$)
$F(\cdot)$	numerator polynomial of output feedback
$\mathcal{H}_i$	i-th hypothesis
I	information sequence
$\mathsf{I}(\cdot)$	right continuous unit step function
k	discrete time variable
L	terminal decision loss

$\mathcal{L}_k$	log likelihood ratio
$m(\cdot)$	moment generating function of z_k
M	number of hypotheses
$M_k(\cdot)$	moment generating function of $\mathcal{L}_k$
n	sample number of SPRT
n_a	degree of A
n_f	degree of F
$\bar{n}$	run length of CUSUM test
$p_k(\theta_i)$	a posteriori probability of $\mathcal{H}_i$
$P_\alpha(\cdot)$	OC function
$P(\cdot)$	denominator polynomial of output feedback
q^{-1}	backwards shift operator
r_K	minimum conditional Bayes risk for the truncated problem
$\bar{R}$	Bayes risk
$\mathbf{R}$	set of real numbers
$\mathcal{S}_k$	cumulative sum
y	system output or observed data
$\mathbf{y}$	vector of outputs
Y	sequence of observed data
u	system input
U	sequence of inputs
$\mathcal{U}$	set of admissible inputs
z_k	increments of $\mathcal{L}_k$
α	lower threshold of SPRT
β	upper threshold of SPRT
$\bar{\beta}\ (\bar{\beta}_{ij})$	threshold(s) of (multihypothesis) CUSUM test
β_{ij}	thresholds of multihypothesis SPRT

γ	input function
Γ	input law
δ_i	terminal decision in favour of $\mathcal{H}_i$
Δ	set of terminal decisions
ϵ	white system noise
ϵ_1 (ϵ_2)	probability of Type I(II) error
$\xi(\cdot)$	one sided spectral distribution function of u
σ	standard deviation of ϵ
φ	regressor vector
ϕ	stopping function
Φ	stopping rule
ψ	stopping function
Ψ	stopping rule
θ_i ($\boldsymbol{\theta}_i$)	parameter(s) describing i-th hypothesis
Θ	set of states of nature
ω	frequency

List of Acronyms

ADD	average detection delay
AR	autoregressive
ARL	average run length
ARMA	autoregressive moving average
ASN	average sample number
CARMA	controlled autoregressive moving average
CUSUM	cumulative sum
FAC	false alarm constraint
GLR	generalized likelihood ratio
GSPRT	generalized sequential probability ratio test

i.i.d.	independent and identically distributed
IPC	input power constraint
MTBFA	mean time between false alarms
OC	operating characteristic
SPRT	sequential probability ratio test

Chapter 1
Introduction

This work is concerned with the detection of abrupt changes in dynamical systems. It deals mainly with statistical techniques and aims to develop input design methods to improve their performances.

A practical way to characterize different possible operation modes of behaviour in a system is to assume a fixed parametrized structure and to describe them via the parameters. In this work, the modes corresponding to possible changes of interest (or the *hypotheses* concerning the system) are assumed to be known. Most of the effort is directed towards the *small change* case where these hypotheses are close in some sense. Naturally, this is where the relevance of the change detection problem is most evident, since in most practical situations big changes would reveal themselves in a catastrophic way.

1.1 Change Detection

The problem of detecting abrupt changes in dynamical systems has gained importance recently as the demand for fault tolerant and reliable engineering systems has increased. The detection of malfunctions and performance degradations in complex automatic systems is crucial to assure safe and low cost operation. In critical applications such as aircraft flight or nuclear power plant control systems it is of vital importance in avoiding hazards to personnel. This aspect of change detection is referred to by many authors as *fault detection and diagnosis* or *instrument fault detection*. The theory and many applications have been covered in books by Patton *et al.* (1989), Basseville and Benveniste (1986), Pau (1981) and Himmelblau (1978) and also in survey papers by Basseville (1988), Gertler (1988), Minorovskii (1981), Isermann (1984) and Willsky (1976).

Even if a change in the system does not necessarily mean an instrument fault, change detection and diagnosis techniques may still be needed in cases such as adaptive control schemes which involve nonsmooth changes of the operation point of the plant or tracking of manoeuvring vehicles. Change detection methods also have applications in the signal processing area. They are widely used in the segmentation of speech and seismic signals (Basseville and Benveniste, 1986, Ch. 3, 10–12; 1983) and image processing (Therrien, *et al.*, 1986).

As shown in Figure 1.1, the two main issues in the design of change monitoring systems are the *generation of signals* or *residuals*, highlighting the possible changes of interest, using the measurements from the process, and *decision making* based on these generated data. Some of the main approaches to these problems are briefly summarized below.

1.1.1 Generation of residuals

The earliest methods for generating change indicating signals were based on hardware redundancy. In this approach one has (at least) three identical instruments. Then a change in any one of them can be detected by comparing the signals obtained from these instruments, thus providing fault tolerant operation (Willsky, 1976). This method can provide fast detection of big changes. However, it can only be implemented in systems where the design and cost limitations leave room for high parallel redundancy. Another drawback of this approach is that, since identical instruments have to be used in parallel, they might develop some faults simultaneously due to a particular reason. Also their expected life of fault-free operation will be alike; so, when one of them malfunctions the others might be expected to do so as well.

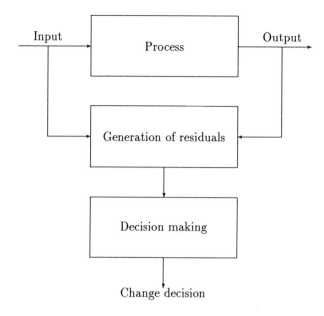

Figure 1.1: Basic stages in a change detection scheme

These limitations together with developments in the space and cost efficient implementation of microcomputers have accounted for the increasing use of *software* or *analytical* redundancy techniques in the generation of residuals. Such techniques are based on processing measurements of dissimilar sensors. Although the instruments are dissimilar they are driven by the same process and, hence, are functionally related. These functional relationships can be exploited to isolate a change in any one of the instruments.

The main concern in residual generation using analytical redundancy techniques is robustness. That is, the residuals are required to signify the possible changes of interest only, and be insensitive to other changes such as external disturbances, set point changes or model uncertainties. Considerable work on the generation of robust residuals has been done by Chow and Willsky (1984a) and Lou *et al.* (1986).

In generating robust residuals many authors have considered using Luenberger

Chapter 1. Introduction   4

observers (Clark, 1978; Clark and Setzer, 1980; Frank and Keller, 1980; Patton *et al.* 1989; Ch. 1–4). The state estimates obtained from several observers can be used in a voting logic to distinguish a faulty instrument. Patton and Kangethe (1989) and Patton and Chen (1991) have investigated the problem of assigning an eigenstructure to these observers in order to make them insensitive to external inputs and disturbances. Recently, Frank *et al.* (1991) have given the necessary conditions to obtain residuals via observers which are totally insensitive to nuisance inputs and disturbances and have discussed optimal compromises whenever this is not possible.

In many cases where the measurements are corrupted by noise, Kalman filters can be employed instead of observers. Newbold and Ho (1968), Willsky and Jones (1976) and Brumback and Srinath (1987) have investigated Kalman filtering methods for change detection.

Both Kalman filtering and observer design techniques seek information about the occurrence of a change in state estimates. Further, the parameter estimates obtained from measured data can also serve as change indicating quantities. For example, Chen and Norton (1987) have used lowpass filtered updates of parameter estimates to detect abrupt changes in system dynamics. Parameter estimation techniques can be used with input-output models, which may be computationally more advantageous than dealing with state space models. A survey of change detection methods via parameter estimation is given by Isermann (1984). Homssi and Despujols (1991) have addressed problems related to identifiability of physical process coefficients, and Geiger (1984) and Baskiotis *et al.* (1979) have reported some applications. As will be discussed below, parameter estimation techniques may also be needed in the calculation of decision statistics.

1.1.2 Decision making

After generating residuals which contain information about the change, the next step is to use them to make a decision on the occurrence and type of change. In most

Chapter 1. Introduction

cases, one is faced with the problem of online detection of the change. Therefore, an important aspect of decision making in change detection problems is its sequential nature. The decision procedure must be able to incorporate the incoming data in an efficient recursive way.

One of the main goals at this step is to detect the change as fast as possible after it has occurred. On the other hand, one would also like to avoid false alarms as long as there is no change in the system. These requirements are usually in conflict, as a detection mechanism responding very quickly to abrupt changes will be, unavoidably, sensitive to the noise corrupting the data. Hence, most decision methods ensure a compromise between detection and false alarm performances, usually by optimizing one of them while guaranteeing a specified level on the other.

The decision making in change detection systems generally involves statistical tests, i.e., checking certain statistics against thresholds. Once the different modes of operation, in which a system possibly can be found, are specified by parametrized models, a set of such statistics can be the a posteriori probability of each model being valid. This method is called the multiple model approach and discussed in detail by Willsky (1986). He considered obtaining the a posteriori probability of each model recursively by using innovations from a bank of Kalman filters, one based on each of the possible models. These probabilities are then compared to prespecified thresholds to determine which mode the system is in and, hence, to detect the occurrence and type of change. Some applications of this approach can be found in articles by Gustafson *et al.* (1978) and Willsky *et al.* (1980).

Another way to compare the likelihood of two different models is by updating the *ratio* between their probabilities at each sampling instant. The *probability* (or *likelihood*) *ratio* can then be compared to two thresholds to determine the valid model. This leads to the so called *sequential probability ratio test* (SPRT). SPRT was originally proposed and investigated in detail by Wald (1947). It has been one of the most frequently discussed topics in sequential decision theory since then (see, e.g., Ghosh, 1970 and references therein). Apart from its simplicity, its optimality property is also one of the main reasons for the special attention it has received: Wald and Wolfowitz (1948) have shown that SPRT, among all sequential tests with given error probabilities, minimizes the average number of data samples required to reach a decision, if the samples are distributed identically and independently.

Several authors have considered modifications of SPRT for purposes of change detection in dynamical systems (Newbold and Ho, 1968; Chien and Adams, 1976; Deckert et al., 1977; Zhang, 1989). Note that SPRT is a test to decide which of the two models is generating all the observed data, rather than whether there is a change in the regime generating the data. Therefore, in using SPRT for change detection purposes it must be applied in a modified way. For example, Chien and Adams (1976) and Zhang (1989) considered resetting the likelihood ratio to some suitable lower bound to reduce its delay in reacting to a change.

The statistical procedure thus obtained is known as the *cumulative sum* (CUSUM) *algorithm*. It was originally proposed by Page (1954). Its performance and optimality properties in detecting changes in a sequence of independently distributed random variables have been discussed by Lorden (1971) and Shiryaev (1978). Nikiforov (1986) considered some modifications of the CUSUM algorithm for the detection of changes in cases where the hypothesis after the change is not known exactly.

The SPRT and CUSUM techniques are basically tests concerning two possible hypotheses about the data. However, they can be extended to test multihypotheses situations and suitable algorithms have been given by Sobel and Wald (1949), Armitage (1950) and Fleisher and Shwedyk (1980). They are based on running SPRT's between several hypothesis pairs in parallel. Recently, Zhang (1989) combined Armitage's SPRT with a resetting rule to obtain a change detection algorithm for more than two hypotheses.

The SPRT and CUSUM algorithms are simple to implement and yet efficient in testing close hypotheses or detecting small changes. In this work, they are investigated in detail.

The SPRT requires exact knowledge of the hypotheses in the calculation of the test statistics. Nevertheless, estimation techniques can be incorporated into the decision procedure to detect changes towards unknown hypotheses. Such tests are known as *generalized likelihood ratio* (GLR) tests (Willsky and Jones, 1976; Willsky, 1986). The basic idea behind GLR is to replace the unknown change time and parameters describing the hypothesis after the change by their maximum likelihood estimates based on the observed data. This is generally feasible when dealing with additive biases in the noise rather than changes in the dynamics. Even in this case, the maximization with respect to the change time can be quite elaborate, necessitating simplifying assumptions (Willsky and Jones, 1976). Various applications of the GLR method are given by Gustafson *et al.* (1978), Willsky *et al.* (1980) and Dowdle *et al.* (1983).

As one is confronted in change detection with a sequential decision problem, its formulation in a Bayesian framework using decision theoretic tools can be also useful in characterizing optimal decision rules. Such an approach has been pursued by Chow and Willsky (1984b). It has the advantages and disadvantages common to most Bayesian approaches in many problems: In practical cases, the optimal decision procedure may be too complex to be implemented. However, it can provide a standard against which various practical methods can be compared.

1.2 Input Design

The inputs required to achieve desired operating conditions in a dynamical system may be designed prior to and independent of the change monitoring system. Such inputs, as well as other disturbances which are not controlled variables, are usually considered as nuisance signals as far as the change detection mechanisms are concerned. As mentioned in the previous section, several authors have investigated the problem of obtaining residuals and change detection methods insensitive to such inputs.

However, if an input (even a small perturbation) is at one's disposal solely for improving the change detection performance, then the problem becomes that of finding input signals to which the detection method is most sensitive, rather than devising detection algorithms which are insensitive to these inputs. Evidently, a good choice of input should improve the performance of the detection algorithm.

The design of optimal inputs has been one of the main research areas in the system identification context. A detailed treatment of the problem has been given in the books by Goodwin and Payne (1977), Zarrop (1979), Kalaba and Spingarn (1982) and has been discussed in numerous articles. (e.g., Mehra, 1974; Uphadyaya and Sorenson, 1977 and Yuan and Ljung, 1984). Much of the effort on input design for system identification has been put into maximizing various cost functions related to the information matrix of the model parameter estimates in order to get greater accuracy.

Some work on input design for model discrimination has been done by Uosaki *et al.* (1984) and Uosaki and Hatanaka (1987). They have discussed design of optimal offline input signals to determine the order of autoregressive models under various constraints. The design was aimed at increasing a measure of distance between two hypothesized models to facilitate better discrimination.

The literature on input design for change detection is relatively sparse and new (Zhang, 1989; Zhang and Zarrop, 1988; Kerestecioğlu and Zarrop, 1989, 1991). Zhang (1989) has discussed both offline designs and online algorithms to generate input signals. She has put more emphasis on controlled autoregressive moving average models and presented applications on chemical processes. The only design criterion considered was the average delay in detecting the change.

However, the false alarm rate is another important factor in assessing the detection performance and, hence, needs to be taken explicitly into account in the input design. This point of view is adopted in developing input design methods in this work.

1.3 Review and Original Contributions

The basic strategy in this work can be outlined as follows: To start with, the *combined* optimal design of detection mechanism and inputs is considered from the viewpoint of Bayesian decision theory (Chapter 2). This methodology leads usually to quite complicated optimal designs and may turn out to be computationally infeasible. Naturally, in choosing a statistical test with related thresholds, simplicity of its implementation plays a major role. Nevertheless, once this choice has been made, one would still have the option of using the best inputs in order to get better performance out of a possibly suboptimal test. Therefore, with this idea of separation in mind, the effort is first directed towards the investigation of some statistical tests (Chapters 3–5) and then to the design of inputs to improve their performances (Chapter 6). The analyses are given in somewhat more detail for the case where one has only two hypotheses concerning the system. However, the results obtained are extended to the multihypothesis case as well (Chapter 7).

Chapter 2 defines the elements of sequential decision theory (Ferguson, 1967), which can be used to obtain optimal decision rules for change detection, in such a way as to accommodate also input design. The decision rule is augmented by an input design rule in Section 2. The optimal decision rules are analyzed using a dynamic programming approach (Section 3). Theorems 2.1–4 are extensions to the classical case (where no input is available to the decision maker) and show that some important properties of the decision rules still hold. In Section 4, special emphasis is put on a two-hypotheses case which also covers the possibility that the data-generating mechanism obeys a regression model (i.e. the data are not independent). It is shown that the optimal statistical test in this case is structurally the same as SPRT save that the thresholds depend on the most recently obtained data. This description of the optimal test for the parameters of dynamical models is believed by the author to be novel. It also gives support to the use of SPRT as a suboptimal test.

Chapter 3 is a survey of some basic results related to SPRT. This chapter deals with the case where the observations are identically and independently distributed. The performance measures of SPRT, namely the average sample number (ASN) to reach a decision and the operating characteristics (OC) are considered (Section 2). A fundamental identity involving the moment generating function of the increments of the log likelihood ratio is introduced, from which these measures can be obtained easily.

The analysis of Chapter 3 is extended in Chapter 4 to the case where SPRT concerns a decision about the parameters of a scalar stationary autoregressive process. Although the final results in this chapter could have been pulled out as special cases of rather general results about the properties of the likelihood ratio due to Eisenberg and Ghosh (1979), it is preferable and more illuminating to construct them using the concepts of the particular problem at hand. In this sense, it is this method of synthesis, leading first to the fundamental identity and then to the ASN and OC formulae, which is original rather than these results themselves. The key lemma of the development is Lemma 4.1 which is about the asymptotic behaviour of the conditional log likelihood ratio (Section 2). The function $\lambda(t)$ introduced in Lemma 4.1 bears properties similar to that of the moment generating function of the increments of the log likelihood ratio in the i.i.d. case. These properties are given in Lemma 4.2. These lemmas lead to an analogue of the Fundamental Identity for the autoregressive case (Section 3). This, in turn, is used to obtain the ASN and OC functions in Section 4. Some points of the analysis which are different in the exact log likelihood ratio case are also indicated (Section 5).

The use of SPRT for change detection purposes is considered in Chapter 5. It introduces the CUSUM algorithm as an SPRT combined with a resetting rule to reduce the detection delay (Section 1). The application to controlled autoregressive moving average models is also discussed. In this chapter, average run length of the test turns out to be an important quantity by which the detection and false alarm performances can be characterized (Section 2). The formulae for average run length under either hypothesis are derived.

The material in Chapter 6 is new and deals with input design. The chosen design objective is to minimize the detection delay in the CUSUM test while ensuring a tolerable mean time between false alarms (Section 2). Both offline and online inputs are considered in achieving this goal. In the offline case (Section 3), it is shown that the classical tradeoff in the change detection problem holds also from the input design point of view: Any input which improves the detection delay is bound to deteriorate the false alarm performance. Theorem 4.1 describes the optimal inputs of constrained power as consisting of at most two frequencies. In Section 3, the design method for finding the optimal frequencies is also described. The online inputs (Section 4) are considered to be generated by linear output feedback with a constraint on either the output or the input power. The optimal solution for a general feedback law has not been possible. However, a suboptimal solution is obtained for a polynomial feedback under output power constraint by linearizing the cost and constraint functions of the related optimization problem.

Chapter 7 is a generalization of some results in the previous chapters to the multihypothesis case. Different extensions of SPRT due to Sobel and Wald (1949) and Armitage (1950) are mentioned in Section 1. More emphasis is put on Armitage's multihypothesis SPRT and it is used to obtain a new CUSUM procedure to detect a change towards one of several hypotheses (Section 2). This new scheme is contrasted with another change detection procedure proposed by Zhang (1989) and also based on Armitage's SPRT. In Section 3, the input design methods of Chapter 6 are extended to the multihypothesis case. In the offline case, the optimal input spectrum is shown to have at most as many frequencies as the number of hypotheses and a

Chapter 1. Introduction 12

search procedure is described to obtain the optimal input. A possible extension to the online input case is also indicated.

Chapter 2

A Decision Theoretic Approach to Change Detection

The Bayesian decision theory provides us with useful tools for formulating the change detection problem as a sequential decision problem. In this chapter we will discuss how to merge the problems of determining optimal inputs and taking optimal decisions about the occurrence of the change within this framework.

First we will review some difficulties in the application of decision theoretic methods to the change detection and diagnosis problem and briefly discuss how M-ary hypothesis testing techniques can provide a useful starting point (Section 1). The elements of a sequential decision problem and the decision rule which take the input available to the decision-maker into account are presented in Section 2. Section 3 discusses the solution for the optimal decision rule. It states and proves some of the basic theorems of decision theory within this augmented viewpoint. Towards this goal, firstly, a truncated decision problem is discussed and then it is shown that the untruncated problem can be approximated asymptotically. In Section 4, we analyze a case where the statistical dependence of the observations is of a special form; namely, where each observation depends explicitly only on m previous observations. After an example given in Section 5 to illustrate the ideas presented, Section 6 consists of some conclusions.

Chapter 2. A Decision Theoretic Approach 14

2.1 Introduction

The problem of determining the decision rules for change detection can be treated as a subclass of the sequential decision problems on which a wide literature has been accumulated (Blackwell and Girshick, 1954; DeGroot, 1970; Ferguson, 1967). A Bayesian approach has been developed by Chow and Willsky (1984b) to solve the decision making problem in change detection. Although the optimal decision rules may be practically uncomputable in many cases, the decision theoretic formulation of the problem provides a conceptual framework and facilitates inclusion of many factors and trade-offs (such as speed of detection, false alarm rate, diagnosis accuracy, etc.) in the analysis of the problem.

In general, a sequential decision rule specifies to the statistician, (i) whether to continue or stop collecting data (stopping rule) and (ii) which decision to take after stopping (terminal decision rule) (Ferguson, 1967). If the data generator under consideration has an input available sequentially to the statistician, which may be the case for a dynamical system under change monitoring, it should tell also (iii) which input to apply if one has decided to continue collecting data. In other words, the stopping rule has to be augmented by an input design rule. In this case, the sequential decision problem involves not only the design of optimal stopping and terminal decision rules, as aimed for in classical decision theory, but also choosing the optimal input sequence which minimizes a given risk function. Below we will analyze the change detection problem from this extended point of view.

As reported by Chow and Willsky (1984b), various aspects peculiar to the change detection and diagnosis problem make the optimal decision rules practically uncomputable. For instance, if an attempt is made to identify both the onset time and type of change simultaneously, one would be confronted with an increasing number of possible change times as the collection of data proceeds. Therefore proper simplifications and assumptions are unavoidable to make the analysis and computation of optimal rules feasible.

One way of dealing with the exponentially increasing number of hypotheses is to assume some onset time for the change and to identify the possible change which may have occurred at that time. That is, a general change detection and diagnosis problem is approximated by a diagnosis problem. Once the decision rules are determined for this type of problem a suboptimal solution for the general one can be obtained in a number of ways.

Chow and Willsky (1984b) used a sliding window approach for that purpose. They assumed that a change may have occurred at the first sampling instant within the data window and applied optimal decision rules to the data within this window to detect and identify the change. This detection method is then implemented by sliding the window forward as the monitoring goes on.

On the other hand, Newbold and Ho (1968) introduced the idea of applying sequentially statistical tests which are designed to distinguish between two hypotheses. According to their strategy, a statistical sequential test is applied to determine whether the system is in abnormal or normal mode. If the test results in the acceptance of the normal mode, another test is started and this resetting is carried out until an abnormal mode is declared.

With these considerations in mind, and for simplicity, we will analyze below the problem of determining the decision rules for detecting changes which may have occurred at a given time. In the decision theoretic context, this will be an M-ary hypotheses testing problem.

2.2 Decision Rules Including Input Law

2.2.1 Elements of sequential decision problem

First, we introduce the elements of a sequential decision problem where an input to the data generating mechanism is available to the statistician:

1) Θ: The set of states of nature. For change detection, Θ is a discrete and finite set whose elements correspond to the normal or an abnormal mode. In most

applications, each θ_i, ($i = 0, \ldots, M-1$), consists of the parameters of a parametrized model describing the operation of the system under different conditions. Various levels of severity of a certain type of change can be represented as different change types at the expense of increasing M. We will denote the a posteriori probability of θ_i being true based on the data collected up to sampling instant k as $p_k(\theta_i)$, and $p_0(\theta_i)$ will be the a priori probability of θ_i before one starts sampling.

2) Δ: The set of terminal decisions taken when one stops collecting data. Each $\delta \in \Delta$ denotes a declaration for a normal or type of abnormal condition. In some cases (Chow and Willsky, 1984b) the normal condition may be absent if the monitoring is to be stopped only to declare a change.

3) $L(\theta, \delta)$: Loss due to making terminal decision δ if the true state of nature is θ. It is assumed that the terminal decision loss is bounded and nonnegative. We will denote the loss of deciding for the j-th type of change if the system is actually in the i-th mode of operation as $L(i, j)$. In particular, the no change condition can be regarded (say) as the 0-th type of change. Hence, $L(0, j)$ is the loss due to a false alarm. An additional property of $L(i, j)$ is

$$L(i,j) > L(i,i) \quad \forall i, \quad \forall j \neq i$$

which means correct diagnosis of the change has less terminal loss than a wrong one.

4) $Y(k) = \{y(i)\}_{i=1}^{k}$: Sequence of observations made up to time k. Each $y(k)$ is a random variable whose distribution may depend on the past observations $Y(k-1)$. They can be the outputs of the monitored system as well as residuals which are obtained by filtering these outputs to highlight different types of changes.

5) $U(k) = \{u(i)\}_{i=0}^{k}$: Sequence of inputs applied up to time k. Each input $u(k)$ is to be chosen from an input space $\mathcal{U}(k)$ which is determined by the constraints imposed on the inputs. We will assume that $\mathcal{U}(k) = \mathcal{U}$, i.e. it is independent of time. Further,

$$I(k) = \{Y(k), U(k-1)\}$$

will represent the information gathered from the system up to time k. The observations are affected by the inputs in a causal way, and the conditional probability density of $y(k)$, given $I(k-1)$, $u(k)$ and that the system is under the i-th type of operation mode is denoted by $f_i(y(k) \mid I(k-1), u(k))$, whereas we will drop the subscript i if no conditioning on the operation mode is meant.

6) $c(k, i, U(k-1))$: Cost of inputs and taking observations up to time k when the true state of nature is θ_i. We assume that it is positive and

$$c(k, i, U(k-1)) < c(k+1, i, U(k)), \qquad (2.1)$$

$$\lim_{k \to \infty} c(k, i, U(k-1)) = \infty.$$

In many practical situations the cost may be separable in $U(k-1)$ and i,

$$c(k, i, U(k-1)) = c_c(k, i) + c_u(k, U(k-1)).$$

That means the cost incurred due to the inputs is independent of the mode of operation of the system. Typical examples are

$$c_c(k, i) = Ck,$$

constant cost per observation;

$$c_u(k, U(k-1)) = 0 \qquad \mathcal{U} = \{u \mid |u| \le K_1\}$$

which represents only an amplitude constraint on $u(k)$ or

$$c_u(k, U(k-1)) = K_2 \sum_{i=0}^{k-1} u^2(i) \qquad \mathcal{U} = \mathbb{R}$$

which represents a power penalty on the input.

2.2.2 Decision rules

Next we will describe how the sequential decision rule should be modified in order to exploit the inputs available to the decision-maker. It can be divided into three main parts.

Stopping rule: The stopping rule is given by the sequence

$$\Phi = \{\phi(k, I(k))\}_{k=0}^{\infty} \tag{2.2}$$

which specifies whether one should stop taking observations to make a decision or whether it is worth taking another observation. The stopping function $\phi(k, I(k))$ is given by

$$\phi(k, I(k)) = \begin{cases} 1 & \text{if sampling is terminated at time } k \text{ given that} \\ & \text{it is not terminated up to time } k \text{ and } I(k) \text{ is} \\ & \text{observed,} \\ 0 & \text{otherwise.} \end{cases}$$

The stopping rule can also be expressed as

$$\Psi = \{\psi(k, I(k))\}_{k=0}^{\infty} \tag{2.3}$$

where

$$\psi(k, I(k)) = \begin{cases} 1 & \text{if sampling is not terminated up to time } k \\ & \text{and it is terminated at time } k \text{ given that } I(k) \\ & \text{is observed,} \\ 0 & \text{otherwise.} \end{cases}$$

Clearly, the difference between $\phi(k, I(k))$ and $\psi(k, I(k))$ lies in the fact that the former describes the stopping at the k-th sampling instant conditional upon that the test is conducted up to then, whereas no such condition is used in the latter. The relation between these two functions can be expressed as

$$\psi(k, I(k)) = \phi(k, I(k)) \prod_{i=0}^{k-1} (1 - \phi(i, I(i))), \qquad k > 0$$

with $\psi(0) = \phi(0)$.

If the true state of nature is θ_i, then the probability of stopping at time k is $E\{\psi(k, I(k)) \mid \theta_i\}$. The probability that the decision procedure will eventually reach a terminal decision is assumed to be 1,

$$\sum_{k=0}^{\infty} E\{\psi(k, I(k)) \mid \theta_i\} = 1, \tag{2.4}$$

Chapter 2. A Decision Theoretic Approach

so that one has a finite loss plus cost under every possible state of nature.

Terminal decision rule: This is another sequence

$$D = \{d(k, I(k))\}_{k=0}^{\infty} \qquad (2.5)$$

where $d(k, I(k))$ is a function which maps the information $I(k)$ into the decision space Δ and indicates the terminal decision taken when one ceases taking observations. Since the terminal decision is to be made only if one stops sampling, $d(k, I(k))$ need be given only for values of k and $I(k)$ where $\psi(k, I(k)) = 1$.

In conventional decision theory, D and Ψ (or Φ) would constitute the decision rule for a sequential decision problem. In the case, however, where an input to the data generating mechanism is available to the statistician, one will have an additional component in the decision rule.

Input law: This is denoted by

$$\Gamma = \{\gamma(k, I(k))\}_{k=0}^{\infty} \qquad (2.6)$$

where $\gamma(k, I(k))$ is a function which maps $I(k)$ into the input space $\mathcal{U}$ and specifies the input to be applied if one decides to continue taking observations. Note that the values of $\gamma(k, I(k))$ are irrelevant if $\psi(i, I(i)) = 1$ for some $i \leq k$ and $I(k) = \{I(i), y(i+1), \ldots, y(k), u(i), \ldots, u(k-1)\}$, because the input is to be applied only if one decides to take more observations.

In other words, the sequential decision rule should tell the decision-maker not only *whether* to continue sampling but also *how* to continue if it is worth making another observation. Hence, the stopping rule together with the input law can be regarded as an *augmented stopping rule*.

Finally, note that the input law in (2.6) as well as other components of the decision rule given in (2.2) (or (2.3)) and (2.5) are nonrandomized (*pure*) decision rules (DeGroot, 1970). Although, for simplicity, we will carry on with this type of decision rule in the forthcoming sections, there is no reason why such rules cannot be replaced by randomized ones. In that case, $\gamma(k, I(k))$ as well as $\psi(k, I(k))$ (or

$\phi(k,I(k)))$ and $d(k,I(k))$ would be interpreted as probability density functions on $\mathcal{U}$, $\{0,1\}$ and Δ, respectively.

2.2.3 Bayes risk

If θ_i is the true state of nature and the sequential decision rule (Φ, Γ, D) is used, then total expected cost plus loss (or the *risk function*) is given by (Ferguson, 1967)

$$R(i, \Phi, \Gamma, D) = \sum_{k=0}^{\infty} E\left\{\psi(k, I(k))\left[c(k, i, U(k-1)) + L(i, d(k, I(k)))\right]\right\}. \quad (2.7)$$

Note that the expectation in (2.7) is over $I(k)$ under the state of nature specified by i.

A Bayesian decision rule (Φ^*, D^*, Γ^*) is one which minimizes the *Bayes risk*

$$\begin{aligned}
\bar{R}(\Phi, \Gamma, D) &= E\{R(i, \Phi, \Gamma, D)\} \\
&= \sum_{i=0}^{M-1} p_0(\theta_i) R(i, \Phi, \Gamma, D) \quad (2.8)
\end{aligned}$$

where $p_0(\theta_i)$ is a probability density over Θ which represents the a priori information about the possible occurrence of the change.

2.3 Dynamic Programming Solution for the Decision Rules

The derivation of a Bayesian decision rule can be carried out in two stages. First we will assume a fixed augmented stopping rule (Φ, Γ) and find the corresponding optimal terminal decision rule. Then, we will try to determine the optimal augmented stopping rule for the derived optimal terminal decision rule.

2.3.1 Optimal terminal decision rule

The optimal terminal decision rule turns out to be the same as in the case where no input is available to the statistician (Ferguson, 1967). We define the minimum conditional expected loss as

$$L^*(k, I(k)) = \min_d E\{L(i, d(k, I(k))) \mid I(k)\}$$

Chapter 2. A Decision Theoretic Approach   21

where the expectation is over different states of nature specified by i and denote $d^*(k, I(k))$ for which the value of $L^*(k, I(k))$ is attained as the Bayes decision rule for fixed sample size. We have the following theorem.

Theorem 2.1 *For any fixed stopping rule and input law, the Bayes risk is minimized by $D^* = \{d^*(k, I(k))\}_{k=0}^{\infty}$.*

Proof. The proof is very similar to the case when the decision rule contains only Φ and D. We present it here for the sake of completeness.

The optimal terminal decision rule minimizes the risk function given in (2.8), namely

$$\bar{R}(\Phi, \Gamma, D) = \sum_{k=0}^{\infty} E\left\{\psi(k, I(k))c(k, i, U(k-1))\right\} + \sum_{k=0}^{\infty} E\left\{\psi(k, I(k))L(i, d(k, I(k)))\right\}$$

where the expectation is taken both over $I(k)$ and θ_i. Since Φ and Γ are fixed, we have to minimize $E\{\psi(k, I(k))\, L(i, d(k, I(k)))\}$ for each k:

$$\min_{d} E\{\psi(k, I(k))\, L(i, d(k, I(k)))\}$$
$$= \min_{d} [E\{\psi(k, I(k))\}\, E\{L(i, d(k, I(k))) \mid I(k)\}]$$
$$= E\{\psi(k, I(k))\} \min_{d} E\{L(i, d(k, I(k))) \mid I(k)\}$$
$$= E\{\psi(k, I(k))\}\, d^*(k, I(k)).$$

Hence, $\{d^*(k, I(k))\}_{k=0}^{\infty}$ minimizes $\bar{R}$. □

Remark. The important point of this theorem is that the terminal decision rule does not depend on the augmented stopping rule. If the statistician is supplied only with the data, $I(k)$, the lack of his knowledge on the stopping rule and input law which is used to collect it will not prevent him from finding an optimal terminal decision rule, because he will act as if he is confronted with a fixed sample size problem. This also implies the existence of a biassed input law similar to the existence of biassed stopping rules when the decision rule is composed of the terminal decision and stopping rules only.

Let us clarify this remark by a simple example.

Example 2.1 Assume that the observations are generated by

$$y(k) = \theta\, u(k-1) + \epsilon(k) \tag{2.9}$$

where $\epsilon(k)$ is a sequence of statistically independent Gaussian random variables with zero mean and variance σ and θ can have a value from the set $\Theta = \{\theta_0, \theta_1\} = \{-1, 1\}$. The input set is given as $\mathcal{U} = \{-1, 0, 1\}$. The decision space will be $\Delta = \{-1, 1\}$, each value denoting a decision in favour of the corresponding value of θ. The terminal decision loss is assumed to be $L(i,j) = 1 - \delta_{ij}$ where δ_{ij} denotes the Kronecker delta. Since

$$E\{L(i, d(k, I(k))) \mid I(k)\} = \begin{cases} p_k(\theta_0) & \text{if } d(k, I(k)) = 1 \\ 1 - p_k(\theta_0) & \text{if } d(k, I(k)) = -1 \end{cases}$$

the optimal terminal decision will look like

$$d^*(k, I(k)) = \begin{cases} 1 & \text{if } p_k(\theta_0) < \dfrac{1}{2} \\ -1 & \text{otherwise.} \end{cases} \tag{2.10}$$

The statistics $p_k(\theta_0)$ can be calculated from $I(k)$ by the Bayes formula

$$p_k(\theta_0) = \frac{p_0(\theta_0)\, f_0(Y(k) \mid U(k-1))}{p_0(\theta_0)\, f_0(Y(k) \mid U(k-1)) + (1 - p_0(\theta_0))\, f_1(Y(k) \mid U(k-1))}. \tag{2.11}$$

By using (2.9) and the normality and independence of $\epsilon(k)$'s in (2.11), the optimal terminal decision can be made according to

$$d^*(k, I(k)) = \begin{cases} 1 & \text{if } \sum_{i=1}^{k} y(i)\, u(i-1) > 0 \\ -1 & \text{otherwise} \end{cases}$$

for the case where $p_0(\theta_0) = p_0(\theta_1) = 1/2$.

Now assume that the data are collected by using the stopping functions $\phi(0) = 0$, $\phi(1) = 0$ and $\phi(2) = 1$, regardless of the value of $I(k)$. That means the statistician chooses to take two and only two observations. Further, assume that the following input functions are adopted in collecting the data:

$$\gamma(0) = 1 \tag{2.12}$$

$$\gamma(1, y(1)) = \begin{cases} 1 & \text{if } y(1) < 0 \\ 0 & \text{otherwise.} \end{cases} \tag{2.13}$$

Chapter 2. A Decision Theoretic Approach

Clearly, the hypotheses $\theta = \theta_1$ will be chosen if $y(1) > 0$ or $y(1) \leq 0$ and $y(2) > -y(1)$. Hence,

$$\begin{aligned}
\Pr(\text{choose } \theta_1) &= \Pr(y(1) > 0) + \Pr(y(1) \leq 0)\Pr(y(2) > -y(1) \mid y(1) \leq 0) \\
&= \frac{1}{2} + \frac{1}{2}\Pr(y(2) > -y(1) \mid y(1) \leq 0) \\
&> \frac{1}{2}.
\end{aligned}$$

Although one starts sampling with equal probabilities for each possible state of nature and assumes the same loss for every wrong terminal decision, under the input described by (2.12) and (2.13) the probability of choosing θ_1 is greater than that of choosing θ_0. Note that, for this example, any input independent of past data would be unbiassed.

2.3.2 Augmented stopping rule: Truncated case

As the second part of the analysis of Bayesian decision rules we will find the optimal stopping rule and input law which minimizes $\bar{R}(\Phi, \Gamma, D^*)$. As the first step in this part a *truncated* problem will be considered. A sequential decision problem is said to be truncated at K if

$$\sum_{k=0}^{K} \psi(k, I(k)) \equiv 1.$$

That is, in a truncated problem if the observations are taken up to and including time K, one is forced to stop and declare a terminal decision.

Let us denote the minimum expected loss plus cost given that the sampling is stopped at time k as $T(k, I(k))$. That means

$$T(k, I(k)) = L^*(k, I(k)) + E\{c(k, i, U(k-1)) \mid I(k)\}. \tag{2.14}$$

At the $(K-1)$-st step we have two choices at hand. One of them is to stop and take the optimal terminal decision in the way described in the previous subsection. In doing this, a cost plus loss given as $T(K-1, I(K-1))$ will be incurred. The other possibility is to apply an input and continue sampling by taking another observation,

Chapter 2. A Decision Theoretic Approach   24

which has the expected loss plus cost $E\{T(K,I(K)) \mid I(K-1), u(K-1)\}$. Therefore, if one decides to take the K-th observation, the optimal input will be

$$u^*(K-1) = \arg\min E\{T(K,I(K)) \mid I(K-1), u(K-1)\}. \quad (2.15)$$

The decision whether to stop or not, however, will be taken by comparing the minimum expected loss plus cost for the K-th step to the loss plus cost at the $(K-1)$-step. So, an optimal stopping rule for the last step can be written as

$$\phi^*(K-1, I(K-1)) = \begin{cases} 1 & \text{if } T(K-1, I(K-1)) \\ & < \min_{u(K-1)} E\{T(K, I(K)) \mid I(K-1), u(K-1)\} \\ 0 & \text{otherwise.} \end{cases} \quad (2.16)$$

Let us denote the minimum conditional Bayes risk given $I(k)$ for the problem truncated at K by $r_K(k, I(k))$. It is seen from (2.16) and (2.14) that

$$r_K(K, I(K)) = T(K, I(K)) \quad (2.17)$$

$$r_K(K-1, I(K-1))$$
$$= \min\left[T(K-1, I(K-1)), \min_{u(K-1)} E\{r_K(K, I(K)) \mid I(K-1), u(K-1)\}\right].$$
$$(2.18)$$

So, inductively, we obtain

$$r_K(k, I(k)) = \min\left[T(k, I(k)), \min_{u(k)} E\{r_K(k+1, I(k+1)) \mid I(k), u(k)\}\right] \quad (2.19)$$

for $k = 0, \ldots, K-1$. The optimal stopping and input functions can be expressed as

$$\phi^*(k, I(k)) = \begin{cases} 1 & \text{if } T(k, I(k)) < \min_{u(k)} E\{r_K(k+1, I(k+1)) \mid I(k), u(k)\} \\ 0 & \text{otherwise,} \end{cases}$$
$$(2.20)$$

$$\gamma^*(k) = \arg\min_{u(k)} E\{r_K(k+1, I(k+1)) \mid I(k), u(k)\} \quad (2.21)$$

for $k = 0, \ldots, K-1$.

Chapter 2. A Decision Theoretic Approach   25

Theorem 2.2 *The terminal decision rule given in Theorem 2.1 together with the augmented stopping rule described in (2.20) and (2.21) is a Bayesian decision rule for the sequential decision problem truncated at K.*

Proof. Let us denote the augmented stopping rule given in (2.20) and (2.21) as Φ_K^* and Γ_K^*, respectively. Let (Φ, Γ, D) be any other decision rule for the truncated problem which minimizes the Bayes risk and where at least one of Φ, Γ or D is different from Φ_K^*, Γ_K^* and D^*, respectively. Assume that $\bar{R}(\Phi, \Gamma, D) \leq \bar{R}(\Phi_K^*, \Gamma_K^*, D^*)$. Since by Theorem 2.1, D^* is independent of the choice of Φ and Γ, we have $\bar{R}(\Phi, \Gamma, D^*) \leq \bar{R}(\Phi, \Gamma, D) \leq \bar{R}(\Phi_K^*, \Gamma_K^*, D^*)$. But, since (Φ_K^*, Γ_K^*) minimizes the Bayes risk when D^* is used, one can only have $\bar{R}(\Phi, \Gamma, D) = \bar{R}(\Phi_K^*, \Gamma_K^*, D^*)$. Hence, $(\Phi_K^*, \Gamma_K^*, D^*)$ minimizes the Bayes risk. □

2.3.3 The untruncated case

Next we will show that the solution of the untruncated problem can be approximated by the solution of the truncated one for large values of K.

Let $\bar{R}^*$ denote the minimum Bayes risk for the untruncated problem. Note that, by virtue of (2.19), $r_0(0) \geq r_1(0) \geq \cdots \geq 0$. Hence, the sequence $r_K(0)$ converges as $K \to \infty$. Therefore, we have to show that $r_K(0) \to \bar{R}^*$ as $K \to \infty$.

Theorem 2.3 *The minimum Bayes risk of the truncated problem can be made arbitrarily close to that of the untruncated one for sufficiently large K; i.e.,*

$$r_K(0) \to \bar{R}^* \quad \text{as} \quad K \to \infty.$$

Proof. Let (Φ^*, Γ^*, D^*) be an optimum decision rule for the untruncated problem. Let Φ_K be Φ^* truncated at K, i.e.,

$$\phi_K(k, I(k)) = \phi^*(k, I(k)) \quad \text{for } k < K,$$
$$\phi_K(K, I(K)) = 1.$$

Chapter 2. A Decision Theoretic Approach

It follows that

$$\bar{R}(\Phi_K, \Gamma^*, D^*) - \bar{R}^*$$
$$= \bar{R}(\Phi_K, \Gamma^*, D^*) - \bar{R}(\Phi^*, \Gamma^*, D^*)$$
$$= E\{\psi_K(K, I(K))[c(K, i, U^*(K-1)) + L(i, d^*(K, I(K)))]\}$$
$$- \sum_{k=K}^{\infty} E\{\psi^*(k, I(k))[c(k, i, U^*(k-1)) + L(i, d^*(k, I(k)))]\} \quad (2.22)$$

where U^* denotes the inputs determined by Γ^*. Note that the set of $I(k)$'s for which $\psi_K(k, I(k)) = 1$ is identical with the set for which $\sum_{k=K}^{\infty} \psi^*(k, I(k)) = 1$, for $k \geq K$. Hence, we can rewrite (2.22) as

$$\bar{R}(\Phi_K, \Gamma^*, D^*) - \bar{R}^*$$
$$= \sum_{k=K}^{\infty} E\{\psi^*(k, I(k))[c(K, i, U^*(K-1)) - c(k, i, U^*(k-1))$$
$$+ L(i, d^*(K, I(K))) - L(i, d^*(k, I(k)))]\}.$$

Using (2.1), we obtain

$$\bar{R}[\Phi_K, \Gamma^*, D^*] - \bar{R}^* \leq \sum_{k=K}^{\infty} E\{\psi^*(k, I(k)) L(i, d^*(K, I(K)))\}$$
$$\leq E\{\psi_K(K, I(K)) L(i, d^*(K, I(K)))\} \quad (2.23)$$

Since the loss, L, is bounded, in view of (2.4), the last term in (2.23) can be made arbitrarily small by increasing K. Therefore,

$$\bar{R}(\Phi_K, \Gamma^*, D^*) \leq \bar{R}^* + \varepsilon$$

for any $\varepsilon > 0$ and sufficiently large K. Since $\bar{R}^* \leq r_K(0) \leq \bar{R}(\Phi_K, \Gamma^*, D^*)$, we conclude $r_K(0) \to \bar{R}^*$ as $K \to \infty$. □

There can be cases where the minimum Bayes risk of the untruncated problem is equal to that of a corresponding truncated problem, rather than being just approximated by it. If the maximum reduction in loss achievable by taking an extra observation is expected to be less than the minimum cost for the next observation beyond a given sampling time, the statistician will never be willing to take any observation thereafter. The next theorem formalizes this idea.

Chapter 2. A Decision Theoretic Approach   27

Theorem 2.4 *If for all $k > K_0$*

$$L^*(k-1, I(k-1)) - \min_{u(k-1)} E\{L^*(k, I(k)) \mid I(k-1), u(k-1)\}$$
$$\leq \min_{u(k-1)} E\{c(k, i, U(k-1)) - c(k-1, i, U(k-2)) \mid I(k-1), u(k-1)\}$$
(2.24)

then $r_{K_0}(0) = \bar{R}^*$.

Proof. From (2.14), we have

$$\min_{u(k-1)} E\{T(k, I(k)) \mid I(k-1), u(k-1)\}$$
$$= \min_{u(k-1)} E\{L^*(k, I(k)) + c(k, i, U(k-1)) \mid I(k-1), u(k-1)\}$$
$$= \min_{u(k-1)} E\{L^*(k, I(k)) \mid I(k-1), u(k-1)\}$$
$$+ \min_{u(k-1)} E\{c(k, i, U(k-1)) \mid I(k-1), u(k-1)\}.$$

By using (2.24), we get

$$\min_{u(k-1)} E\{T(k, I(k)) \mid I(k-1), u(k-1)\}$$
$$\geq L^*(k-1, I(k-1)) + E\{c(k-1, i, U(k-2)) \mid I(k-1)\}$$
$$\geq T(k-1, I(k-1)).$$

Therefore, from (2.17) and (2.19), we obtain by induction $r_K(k) = T(k, I(k))$ for $K \geq k \geq K_0$. This implies $r_K(K_0) = r_{K_0}(K_0)$ for $K > K_0$; hence, $r_K(0) = r_{K_0}(0)$. Finally, we let $K \to \infty$ to obtain $r_{K_0}(0) = \bar{R}^*$. □

2.4 A Special Case

In the previous section we have employed rather general loss and cost functions. An interesting special case which has been quite extensively dealt with in the literature is certainly the two hypotheses case where constant cost per observation and independently and identically distributed observations are assumed (Blackwell and Girshick, 1954; DeGroot, 1970; Bertsekas, 1987). In this case, it is well known that the sequential probability ratio test (SPRT) (Wald, 1947) minimizes the Bayes risk. Let us denote the probability density function of the observation vector $\mathbf{y}_k = [y(1), \ldots, y(k)]^T$ under the hypotheses $\mathcal{H}_0$ and $\mathcal{H}_1$ as $f_0(\mathbf{y}_k)$ and $f_1(\mathbf{y}_k)$, respectively. SPRT is conducted by comparing the logarithm of the likelihood ratio

$$\mathcal{L}_k = \ln \frac{f_1(\mathbf{y}_k)}{f_0(\mathbf{y}_k)} \tag{2.25}$$

to precomputed thresholds α and β, where $\alpha < 0 < \beta$. The stopping and terminal decision functions can be summarized as follows:

$$\begin{array}{ll} \text{Terminate sampling and decide for } \mathcal{H}_0 & \text{if } \mathcal{L}_k \leq \alpha \\ \text{Terminate sampling and decide for } \mathcal{H}_1 & \text{if } \mathcal{L}_k \geq \beta \\ \text{Continue sampling} & \text{otherwise} \end{array} \tag{2.26}$$

where $\mathcal{H}_i$ is the hypothesis that θ_i is the true state of nature.

Another optimal property was proved by Wald and Wolfowitz (1948). It states that for given error probabilities SPRT has the least expected number of samples to reach a terminal decision among all sequential procedures. Again, this optimality of SPRT is valid if the observations are independently and identically distributed.

The properties and performance of SPRT will be extensively reviewed in Chapter 3. In this section, we will be concerned with a two hypotheses case. That is $\Theta = \{\theta_0, \theta_1\}$. We can denote the decision space as $\Delta = \{\delta_0, \delta_1\} = \{0, 1\}$, where δ_i denotes a terminal decision in favour of θ_i.

We will assume that the distribution of each observation depends only on the previous m observations and inputs explicitly. That is,

$$f_i(y(k) \mid I(k-1), u(k-1)) = f_i(y(k) \mid \boldsymbol{\varphi}(k-1), u(k-1)) \qquad i = 0, 1 \tag{2.27}$$

where

$$\boldsymbol{\varphi}(k) = [y(k), u(k-1), \ldots, y(k-m+1), u(k-m)]^T$$

is a regressor vector. Such is the case, for example, if the observations are outputs of an autoregressive moving average process corrupted by white noise.

Chapter 2. A Decision Theoretic Approach

We will further assume that the cost per observation is constant and independent of the inputs, in other words,

$$c(k, i, U(k-1)) = Ck$$

with $C > 0$. Finally, the loss function will be taken as

$$L(i,j) = \begin{cases} 0 & \text{if } i = j \\ L_i & \text{if } i \neq j \end{cases} \qquad i,j = 0,1.$$

Since

$$E\{L(i, d(k, I(k))) \mid I(k)\} = \begin{cases} L_0 p_k(\theta_0) & \text{if } d(k, I(k)) = 1 \\ L_1(1 - p_k(\theta_0)) & \text{if } d(k, I(k)) = 0, \end{cases} \qquad (2.28)$$

the optimal terminal decision will be

$$d^*(k, I(k)) = \begin{cases} 0 & \text{if } p_k(\theta_0) > L_1/(L_0 + L_1) \\ 1 & \text{if } p_k(\theta_0) \leq L_1/(L_0 + L_1). \end{cases} \qquad (2.29)$$

If one stops to take a terminal decision at the k-th instant the minimum expected cost plus loss will be, from (2.14),

$$T(k, I(k)) = \min\left[L_0 p_k(\theta_0), L_1(1 - p_k(\theta_0))\right] + Ck. \qquad (2.30)$$

Note that $T(k, I(k))$ depends on $I(k)$ only through $p_k(\theta_0)$. On the other hand, by (2.27),

$$f(y(k+1) \mid I(k), u(k))$$
$$= p_k(\theta_0) f_0(y(k+1) \mid \varphi(k), u(k)) + (1 - p_k(\theta_0)) f_1(y(k+1) \mid \varphi(k), u(k))$$

Therefore,

$$E\{r_K(k+1, I(k+1)) \mid I(k), u(k)\} = E\{r_K(k+1, I(k+1)) \mid p_k(\theta_0), \varphi(k), u(k)\}.$$

It is seen from (2.19) and (2.30) that $r_K(k, I(k))$ depends on $I(k)$ only through $(p_k(\theta_0), \varphi(k))$. So, $(p_k(\theta_0), \varphi(k))$ is a sufficient statistic for this sequential decision problem. Below, we will find it convenient to replace the argument $I(k)$ of some functions by $(p_k(\theta_0), \varphi(k))$.

Let us first consider the truncated case. Using (2.30) in (2.19), the minimum conditional Bayes risk can be written as

$$r_K(k, p_k(\theta_0), \varphi(k)) = \min\left[L_0\, p_k(\theta_0) + Ck,\, L_1(1 - p_k(\theta_0)) + Ck,\, W_k(p_k(\theta_0), \varphi(k))\right] \tag{2.31}$$

where

$$W_k(p_k(\theta_0), \varphi(k)) = \min_{u(k)} E\{r_K(k+1, p_{k+1}(\theta_0), \varphi(k+1)) \mid p_k(\theta_0), \varphi(k), u(k)\}. \tag{2.32}$$

We have the following lemma.

Lemma 2.1 $W_k(p_k(\theta_0), \varphi(k))$ has the following properties:

i) $W_k(0, \varphi(k)) = W_k(1, \varphi(k)) = C(k+1)$ for all $\varphi(k)$.

ii) For a fixed $\varphi(k) = \varphi$, $W_k(p_k(\theta_0), \varphi(k))$ is a concave function of $p_k(\theta_0)$.

Proof. Since we have nonnegative loss, $r_K(k+1, p_{k+1}(\theta_0), \varphi(k+1)) \geq C(k+1)$; hence, by (2.32), $W_k(p_k(\theta_0), \varphi(k)) \geq C(k+1)$. So, by putting $p_k(\theta_0) = 0$ or $p_k(\theta_0) = 1$ in (2.31) one gets $r_K(k, 0, \varphi(k)) = r_K(k, 1, \varphi(k)) = Ck$. Therefore, $W_k(0, \varphi(k)) = W_k(1, \varphi(k)) = C(k+1)$ for all $\varphi(k)$ and (i) is proven.

To prove (ii), assume that $\varphi(k) = \varphi$. For the last sampling instant K, we have

$$\begin{aligned} r_K(K, p_K(\theta_0), \varphi) &= T(K, p_K(\theta_0), \varphi) \\ &= \min[L_0\, p_K(\theta_0), L_1(1 - p_K(\theta_0))] + CK. \end{aligned}$$

which is a concave function of $p_K(\theta_0)$. On the other hand, in view of (2.31), concavity of $W_k(p_k(\theta_0), \varphi)$ implies the concavity of $r_K(k, p_k(\theta_0), \varphi)$. To complete the proof by induction one needs to show that the concavity of $r_K(k+1, p_{k+1}(\theta_0), \varphi)$ implies that of $W_k(p_k(\theta_0), \varphi)$. Let

$$\eta_1 = p_1\, f_0(y(k+1) \mid \varphi, u(k)) + (1 - p_1)\, f_1(y(k+1) \mid \varphi, u(k)),$$

$$\eta_2 = p_2\, f_0(y(k+1) \mid \varphi, u(k)) + (1 - p_2)\, f_1(y(k+1) \mid \varphi, u(k))$$

Chapter 2. A Decision Theoretic Approach

with $p_1, p_2 \in [0,1]$. Clearly, if $p_k(\theta_0) = p_i$, then $p_i f_0(y(k+1) \mid \boldsymbol{\varphi}, u(k))/\eta_i$ will be the a posteriori probability of θ_0 being true for the case where $\boldsymbol{\varphi}(k) = \boldsymbol{\varphi}$ and the input $u(k)$ is applied at time k; for $i = 0, 1$. From the concavity of $r_K(k+1, p_{k+1}(\theta_0), \boldsymbol{\varphi})$ it follows that

$$\frac{\mu\eta_1}{\mu\eta_1 + (1-\mu)\eta_2} r_K\left(k+1, \frac{p_1 f_0(y(k+1) \mid \boldsymbol{\varphi}, u(k))}{\eta_1}, \bar{\boldsymbol{\varphi}}\right) + $$
$$\frac{(1-\mu)\eta_2}{\mu\eta_1 + (1-\mu)\eta_2} r_K\left(k+1, \frac{p_2 f_0(y(k+1) \mid \boldsymbol{\varphi}, u(k))}{\eta_2}, \bar{\boldsymbol{\varphi}}\right)$$
$$\leq r_K\left(k+1, \frac{(\mu p_1 + (1-\mu)p_2) f_0(y(k+1) \mid \boldsymbol{\varphi}, u(k))}{\mu\eta_1 + (1-\mu)\eta_2}, \bar{\boldsymbol{\varphi}}\right)$$
(2.33)

for all $\mu \in [0,1]$, where $\bar{\boldsymbol{\varphi}}$ is to be taken as to satisfy $[\bar{\boldsymbol{\varphi}}^T, y(k-m-1), u(k-m)]^T = [y(k+1), u(k), \boldsymbol{\varphi}^T]$. Since

$$\mu\eta_1 + (1-\mu)\eta_2 = (\mu p_1 + (1-\mu)p_2) f_0(y(k+1) \mid \boldsymbol{\varphi}, u(k)) + $$
$$(1 - (\mu p_1 + (1-\mu)p_2)) f_1(y(k+1) \mid \boldsymbol{\varphi}, u(k))$$

the inequality (2.33) can be written as

$$\mu\, h(p_1, \boldsymbol{\varphi}, y(k+1), u(k)) + (1-\mu)\, h(p_2, \boldsymbol{\varphi}, y(k+1), u(k))$$
$$\leq h(\mu p_1 + (1-\mu)p_2, \boldsymbol{\varphi}, y(k+1), u(k))$$

where

$$h(p, \boldsymbol{\varphi}, y, u) = [p f_0(y \mid \boldsymbol{\varphi}, u) + (1-p) f_1(y \mid \boldsymbol{\varphi}, u)]$$
$$\times r_K\left(k+1, \frac{p f_0(y \mid \boldsymbol{\varphi}, u)}{p f_0(y \mid \boldsymbol{\varphi}, u) + (1-p) f_1(y \mid \boldsymbol{\varphi}, u)}, \bar{\boldsymbol{\varphi}}\right)$$
$$= f(y \mid \boldsymbol{\varphi}, u) r_K\left(k+1, \frac{p f_0(y \mid \boldsymbol{\varphi}, u)}{f(y \mid \boldsymbol{\varphi}, u)}, \bar{\boldsymbol{\varphi}}\right)$$

So, $h(p, \boldsymbol{\varphi}, y(k+1), u(k))$ is a concave function of p, for any $\boldsymbol{\varphi}$, $y(k+1)$ and $u(k)$. Hence, so is

$$W_k(p, \boldsymbol{\varphi}) = \min_{u(k)} \int_{-\infty}^{\infty} h(p, \boldsymbol{\varphi}, y, u(k))\, dy.$$

□

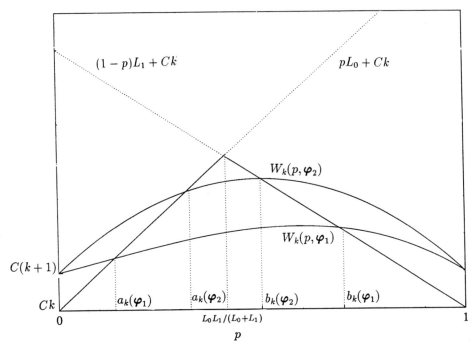

Figure 2.1: The function $W_k(p, \varphi)$ and decision thresholds

Assuming $W_k(p_k(\theta_0), \varphi(k)) < L_0 L_1/(L_0 + L_1)$, it is seen from Figure 2.1 that the optimal decision to stop can be taken by comparing $p_k(\theta_0)$ to two thresholds which depend on the last m observations and inputs, i.e., $\varphi(k)$. Namely,

$$\phi(k, p_k(\theta_0), \varphi(k)) = \begin{cases} 0 & \text{if } a_k(\varphi(k)) < p_k(\theta_0) < b_k(\varphi(k)) \\ 1 & \text{otherwise.} \end{cases}$$

Naturally, if the decision is to take another observation, the input will have the value for which the minimum in (2.32) is attained. The thresholds $a_k(\varphi(k))$ and $b_k(\varphi(k))$ can be related to $W_k(p_k(\theta_0), \varphi(k))$ as

$$W_k(a_k(\varphi(k)), \varphi(k)) = a_k(\varphi(k)) L_0 + Ck, \tag{2.34}$$

$$W_k(b_k(\varphi(k)), \varphi(k)) = (1 - b_k(\varphi(k))) L_1 + Ck. \tag{2.35}$$

Since, by Theorem 2.3, the minimum conditional Bayes risk will converge as $K \to \infty$, for the untruncated problem, one can define the stopping rule with thresholds which do not depend explicitly on k, but only to the data of the last m steps; i.e.,

$$\phi(k, p_k(\theta_0), \boldsymbol{\varphi}(k)) = \begin{cases} 0 & \text{if } a(\boldsymbol{\varphi}(k)) < p_k(\theta_0) < b(\boldsymbol{\varphi}(k)) \\ 1 & \text{otherwise.} \end{cases} \quad (2.36)$$

Finally, by using the Bayes rule,

$$p_k(\theta_0) = \frac{p_0(\theta_0) f_0(\mathbf{y}_k \mid U(k-1))}{p_0(\theta_0) f_0(\mathbf{y}_k \mid U(k-1)) + (1 - p_0(\theta_0)) f_1(\mathbf{y}_k \mid U(k-1))}$$

the stopping rule in (2.36) and the terminal decision rule in (2.29) can be combined into the following decision rule:

$$\text{Terminate sampling and decide for } \mathcal{H}_0 \quad \text{if} \quad \mathcal{L}_k \leq \ln \frac{(1 - b(\boldsymbol{\varphi}(k))) \, p_0(\theta_0)}{b(\boldsymbol{\varphi}(k)) \, (1 - p_0(\theta_0))}$$

$$\text{Terminate sampling and decide for } \mathcal{H}_1 \quad \text{if} \quad \mathcal{L}_k \geq \ln \frac{(1 - a(\boldsymbol{\varphi}(k))) \, p_0(\theta_0)}{a(\boldsymbol{\varphi}(k)) \, (1 - p_0(\theta_0))} \quad (2.37)$$

$$\text{Continue sampling} \quad \quad \quad \quad \text{otherwise}$$

where

$$\mathcal{L}_k = \ln \frac{f_1(\mathbf{y}_k \mid U(k-1))}{f_0(\mathbf{y}_k \mid U(k-1))}.$$

It is seen from (2.37) that the stopping and terminal decision rules put together turn out to be a *quasi*-SPRT where the likelihood ratio is compared to two thresholds which depend on an $2m$-dimensional data vector, at each sampling instant. It is clear from the above analysis, if the observations are independently and identically distributed, the thresholds will be constant; hence (2.37) will reduce to the SPRT in (2.26) (see DeGroot, 1970; Bertsekas, 1987).

It should be noted that this quasi-SPRT is different from the Generalized SPRT, which has been widely investigated in the literature (Ghosh, 1970; Eisenberg, *et al.*, 1976). In GSPRT, the likelihood ratio is compared to thresholds which may assume different values at different sampling instants; nevertheless, they are predetermined for each k, hence, independent of the data. The quasi-SPRT in (2.37) employs thresholds which explicitly depend on the past data.

2.5 An Example

Below a simple example is presented to clarify the optimal decision rules analyzed above.

Example 2.2 Let $\{y(1), y(2), \ldots\}$ be a sequence of samples from a random process which are assumed to be generated by one of the conditional probability functions

$$f_0(y(k) \mid u(k-1)) = \begin{cases} 1 - q - u(k-1) & \text{if } y(k) = 0 \\ q + u(k-1) & \text{if } y(k) = 1, \end{cases} \quad (2.38)$$

$$f_1(y(k) \mid u(k-1)) = \begin{cases} r + u(k-1) & \text{if } y(k) = 1 \\ 1 - r - u(k-1) & \text{if } y(k) = 2, \end{cases} \quad (2.39)$$

where $0 < q < r < 1$ and $u(k)$ is an input satisfying $-q < u_1 \leq u(k) \leq u_2 < 1 - r$. The problem is to find the optimal sequential decision rule for deciding about the conditional distribution of $y(k)$ if the cost and loss functions are given as $c(k, i, U(k-1)) = k$ and

$$L(i, j) = \begin{cases} 0 & \text{if } i = j \\ L & \text{if } i \neq j, \end{cases}$$

respectively. In view of (2.29), the optimal terminal decision will be

$$d^*(k, I(k)) = \begin{cases} 0 & \text{if } p_k(q) > \frac{1}{2} \\ 1 & \text{otherwise} \end{cases}$$

where $p_k(q)$ denoting the probability of q being the true parameter value.

In fact, since the observations are independently distributed, the expected loss plus cost and the minimum conditional Bayes risk at each stage depend on the data only through $p_k(q)$ which, therefore, is a sufficient statistic. By making use of the Bayes rule the evolution of $p_k(q)$ can be described by

$$p_k(q) = \begin{cases} \dfrac{\prod_{i=1}^{k}(q + u(i-1))}{\prod_{i=1}^{k}(q + u(i-1)) + \prod_{i=1}^{k}(r + u(i-1))} & \text{if } y(i) = 1 \;\; \forall i \leq k \\ 1 & \text{if } y(i) = 0 \;\; \text{for some } i \leq k \\ 0 & \text{if } y(i) = 2 \;\; \text{for some } i \leq k \end{cases} \quad (2.40)$$

Since an observation which is equal to 2 or 0 supplies the statistician with perfect knowledge of the true hypothesis, a good stopping rule is not expected to suggest taking any more observations thereafter.

Chapter 2. A Decision Theoretic Approach   35

Let us first make use of Theorem (2.4) to determine if the problem can be solved by a truncated stopping rule. That is, we search for K where for all $k > K$ (2.24) holds. By (2.28) we have

$$L^*(k-1, I(k-1)) = L \min[p_{k-1}(q), 1 - p_{k-1}(q)] \tag{2.41}$$

and

$$\min_{u(k-1)} E\{L^*(k, I(k)) \mid I(k-1), u(k-1)\}$$
$$= \min_{u(k-1)} E\{L \min[p_k(q), 1 - p_k(q)] \mid p_{k-1}(q), u(k-1)\}$$
$$= L \min_{u(k-1)} [\min [p_{k-1}(q)(q + u(k-1)), (1 - p_{k-1}(q))(r + u(k-1))]]$$
$$= L \min [p_{k-1}(q)(q + u_1), (1 - p_{k-1}(q))(r + u_1)]$$
$$\geq (q + u_1) L \min[p_{k-1}(q), 1 - p_{k-1}(q)]. \tag{2.42}$$

By using (2.41) and (2.42), inequality (2.24) will be satisfied if

$$(1 - q - u_1) L \min[p_{k-1}(q), 1 - p_{k-1}(q)] < 1. \tag{2.43}$$

Substituting (2.40) into (2.43) and noting $q < r$ we get

$$(1 - q - u_1) L \frac{\prod_{i=1}^{k-1}(q + u(i-1))}{\prod_{i=1}^{k-1}(q + u(i-1)) + \prod_{i=1}^{k-1}(r + u(i-1))} \leq 1.$$

This inequality is satisfied if

$$(1 - q - u_1)(q + u_2)^{k-1} L \leq (q + u_1)^{k-1} + (r + u_1)^{k-1}$$

or if

$$\left(\frac{q + u_1}{q + u_2}\right)^{k-1} + \left(\frac{r + u_1}{q + u_2}\right)^{k-1} \geq (1 - q - u_1) L. \tag{2.44}$$

If $r + u_1 > q + u_2$, the right hand side can be made arbitrarily large by increasing k, and this guarantees the existence of K such that (2.24) is fulfilled for all $k > K$. For instance, for

$$q = 0.7 \quad r = 0.9 \quad u_1 = -0.05 \quad u_2 = 0.05 \quad L = 10 \tag{2.45}$$

Chapter 2. A Decision Theoretic Approach 36

inequality (2.44) is satisfied for $k > 10$. Hence, given the values in (2.45), it suffices to consider the problem as truncated at $K = 10$ to find the optimal decision rules.

Let us first determine the optimal input law. In view of (2.30), the minimum expected loss plus cost for the k-th sampling instant is

$$T(k, p_k(q)) = L \min[p_k(q), 1 - p_k(q)] + k. \qquad (2.46)$$

At time $K - 1$, the optimal input can be found using (2.15)

$$\begin{aligned} u^*(K-1) &= \arg\min E\{L \min[p_K(q), 1 - p_K(q)] + K \mid p_{K-1}(q), u(K-1)\} \\ &= \arg\min \min \left[p_{K-1}(q)(q + u(K-1)), (1 - p_{K-1}(q))(r + u(K-1)) \right] \\ &= u_1. \end{aligned}$$

By substituting (2.46) and (2.17) into (2.18) and simplifying, the minimum conditional Bayes risk at $K - 1$ turns out to be

$$\begin{aligned} r_K(K-1, p_{K-1}(q)) &= \min \left[L\, p_{K-1}(q), L\, p_{K-1}(q)\,(q + u_1) + 1, L\,(1 - p_{K-1}(q)), \right. \\ &\qquad \left. L\,(1 - p_{K-1}(q))\,(r + u_1) + 1 \right] + K - 1. \end{aligned}$$

In fact, by induction it is straightforward to show that

$$r_K(k, p_k(q))$$
$$= \min \left\{ L p_k(q)(q + u_1)^i + x_i(p_k(q)), L(1 - p_k(q))(r + u_1)^i + x_i(p_k(q)) \right\}_{i=0}^{K-k} + k \qquad (2.47)$$

where

$$x_i(p) = p\,\frac{1 - (q + u_1)^i}{1 - q - u_1} + (1 - p)\,\frac{1 - (r + u_1)^i}{1 - r - u_1}.$$

The input is to be computed at each step by taking the average value of the minimum conditional Bayes risk for the next step, given the information up to then. Taking the expectation of $r(k + 1, p_{k+1}(q))$ in (2.47) and rearranging yields

Chapter 2. A Decision Theoretic Approach   37

$$E\{r(k+1, p_{k+1}(q)) \mid p_k(q), u(k)\}$$
$$= \min\left\{L\,p_k(q)\,(q+u(k))\,(q+u_1)^i + p_k(q)\,(q+u(k))\frac{1-(q+u_1)^i}{1-q-u_1}\right.$$
$$+ (1-p_k(q))\,(r+u(k))\frac{1-(r+u_1)^i}{1-r-u_1},$$
$$L(1-p_k(q))\,(r+u(k))\,(r+u_1)^i + p_k(q)\,(q+u(k))\frac{1-(q+u_1)^i}{1-q-u_1}$$
$$\left.+ (1-p_k(q))\,(r+u(k))\frac{1-(r+u_1)^i}{1-r-u_1}\right\}_{i=0}^{K-k-1} + k+1. \qquad (2.48)$$

It is clear from (2.48) that the expected minimum conditional Bayes risk for the $(k+1)$-st step is minimized by

$$u^*(k) = u_1. \qquad (2.49)$$

Evaluating (2.48) at $u(k) = u_1$ we find

$$W_k(p_k(q))$$
$$= \min\left\{Lp_k(q)(q+u_1)^i + x_i(p_k(q)), L(1-p_k(q))(r+u_1)^i + x_i(p_k(q))\right\}_{i=1}^{K-k} + k. \qquad (2.50)$$

From (2.49), (2.38) and (2.39) it can be seen that the optimal input minimizes the probability of $y(k)$ being one, which is the only possible observation that can cause uncertainty about the state of nature.

To find the optimal stopping rule one has to carry out the comparison described by (2.20). Whenever a 0 or 2 is observed, we have $p_k(q) = 1$ or $p_k(q) = 0$, respectively; hence, no loss will be incurred. In that case the optimal decision would be to stop immediately. On the other hand, the values of $p_k(q)$, $T(k, p_k(q))$, $W_k(p_k(q))$ and $r_K(k, p_k(q))$ corresponding to (2.45) are shown in Table 2.1, for the case where $y(i) = 1$ and $u(k) = u_1$ for all $i \le k$ and $p_0(q) = 0.5$ is assumed. It is seen from Table 2.1 that the expected minimum Bayes risk for the next step is less than the expected loss plus cost up to the 3rd sampling instant. So, the optimal decision will be to take a further observation for $k \le 3$. Thereafter, since $T(k, p_k(q)) < W_k(p_k(q))$,

k	$p_k(q)$	$T(k,p_k(q))$	$W_k(p_k(q))$	$r_K(k,p_k(q))$
0	0.50	5.00	3.66	3.66
1	0.43	5.33	4.55	4.55
2	0.37	5.69	5.34	5.34
3	0.31	6.09	6.01	6.01
4	0.25	6.55	6.66	6.55
5	0.21	7.07	7.35	7.07
6	0.17	7.67	8.08	7.67
7	0.13	8.33	8.86	8.33
8	0.10	9.05	9.68	9.05
9	0.08	9.82	10.53	9.82
10	0.06	10.64		10.64

Table 2.1: Decision variables in Example 2.2 if $y(i) = 1$ for all $i \leq k$

one has to stop at $k = 4$. Therefore, the optimal stopping rule turns out to be

$$\phi(k, I(k)) = \begin{cases} 0 & \text{if } y(k) = 1 \text{ and } k \leq 4 \\ 1 & \text{otherwise.} \end{cases}$$

The stopping rule can also be described in the form of (2.36), i.e., by comparing $p_k(q)$ to two thresholds. Using (2.50) and (2.45) in (2.34) and (2.35), one can compute the thresholds a and b as $a = 0.29$ and $b = 0.63$. Hence, in view of (2.37), the stopping and terminal decision rules can be implemented as

$$\begin{aligned} &\text{Terminate sampling and decide for } \mathcal{H}_0 \text{ if } \mathcal{L}_k \leq \ln\frac{1-b}{b} = -0.543 \\ &\text{Terminate sampling and decide for } \mathcal{H}_1 \text{ if } \mathcal{L}_k \geq \ln\frac{1-a}{a} = 0.916 \\ &\text{Continue sampling} \quad\quad\quad\quad\quad\quad\quad \text{otherwise} \end{aligned} \quad (2.51)$$

where

$$\mathcal{L}_k = \begin{cases} k\ln\left(\dfrac{r+u_1}{q+u_1}\right) & \text{if } y(i) = 1 \quad \forall i \leq k \\ \infty & \text{if } y(i) = 2 \quad \exists i \leq k \\ -\infty & \text{if } y(i) = 0 \quad \exists i \leq k. \end{cases} \quad (2.52)$$

is the log-likelihood ratio between two hypotheses.

2.6 Conclusions

In this chapter, we have shown how the elements of sequential decision theory can be extended to accommodate the design of inputs which are available to the decision-maker. It is shown that if the stopping rule is augmented by an input design rule some important properties of the optimal decision rules are still valid. These include that the optimal terminal decision rule is a collection of fixed sample size Bayes rules and that the problem when the sample size is not restricted can be approximated asymptotically.

We also analyzed a two-hypotheses case where the outputs are statistically dependent in a special way which includes the autoregressive models frequently used for describing linear dynamical systems. It is found that the stopping rule and terminal decision rule can be implemented by monitoring the log likelihood ratio of two hypotheses; a test which reduces to ordinary SPRT if the observations are statistically independent.

The example given in Section 5 is a highly special one. The computation of optimal inputs and decision rules in a more general case is generally quite complicated, if possible at all. Therefore, a suboptimal approach must usually be implemented even for a simple hypothesis testing problem. The results in Section 4 suggest that SPRT can be employed for this purpose, because, although not optimal for the dependent observations case, it is rather effective and easily applicable to the change detection problem in dynamical systems. Therefore, the properties of SPRT will be reviewed extensively in the next chapter.

Chapter 3

The Sequential Probability Ratio Test

A quite large body of theory has grown around the sequential probability ratio test during the last four decades. In this chapter, we summarize some basic results about this test under the assumption that the observations are identically and independently distributed (i.i.d.). Our purpose is to provide a suitable framework which can then later be generalized to deal with dynamical (non-i.i.d.) processes.

The first section introduces the recursive implementation of the log likelihood ratio and explains the relation between the thresholds of the test and its error probabilities. In Section 2, a key identity related to SPRT is quoted and two important quantities characterizing the performance of the test, namely operating characteristics function and average sample number, are derived using it. Two examples are presented in Section 3 and some concluding remarks in Section 4.

3.1 Definition

The sequential probability ratio test is a test to decide between two simple statistical hypotheses. It was investigated originally in depth by Wald (1947). Mainly because of its optimality properties (Wald and Wolfowitz, 1948) and yet its simple implementation, it has been the main subject of *sequential analysis*, the branch of statistics which deals with design, behaviour and performance of sequential statisti-

Chapter 3. The Sequential Probability Ratio Test   41

cal tests. Besides Wald (1947), detailed reviews and analyses of SPRT are given by, e.g., Ghosh (1970), Wetherill and Glazebrook (1986), Siegmund (1985) and Johnson (1961).

The definition of SPRT has already been introduced in Section 2.4. As described in (2.26), at each sampling instant the log likelihood ratio $\mathcal{L}_k$ (see (2.25)) of two hypotheses $\mathcal{H}_0$ and $\mathcal{H}_1$, described by two distinct set of parameters θ_0 and θ_1, respectively, is calculated for as long as

$$\alpha < \mathcal{L}_k < \beta \qquad (3.1)$$

and at the first instant k when (3.1) is violated the test is stopped and a decision is made for $\mathcal{H}_0$ if $\mathcal{L}_k \leq \alpha$ or for $\mathcal{H}_1$ if $\mathcal{L}_k \geq \beta$.

For the case where the observations form a sequence of i.i.d. random variables the log likelihood ratio in (2.25) can be written as

$$\begin{aligned}\mathcal{L}_k &= z_1 + z_2 + \cdots + z_k \\ &= \mathcal{L}_{k-1} + z_k \end{aligned} \qquad (3.2)$$

with $\mathcal{L}_0 = 0$, where

$$z_i = \ln \frac{f_1(y(i))}{f_0(y(i))}. \qquad (3.3)$$

This means that the log likelihood ratio can be computed recursively by adding the scores in (3.3) at each time instant.

The thresholds α and β can be determined to ensure that the probabilities of different types of error are equal to given values. We shall denote Type I error, i.e., the probability of deciding in favour of $\mathcal{H}_1$ when $\mathcal{H}_0$ is true, as ϵ_1 and Type II error, the probability of rejecting $\mathcal{H}_1$ when it is true, as ϵ_2. The quantities ϵ_1 and ϵ_2 are also called *false alarm probability* and *missed alarm probability*, respectively.

Wald (1947) has shown that SPRT will eventually terminate with probability 1 if the observations are independently distributed. For such a case, the probability of making the correct decision will be $1 - \epsilon_1$ if $\mathcal{H}_0$ is true or $1 - \epsilon_2$ if $\mathcal{H}_1$ is true. It

Chapter 3. The Sequential Probability Ratio Test

is also shown by Wald (1947) that the thresholds α and β are related to the error probabilities by

$$\alpha \geq \ln \frac{\epsilon_2}{1-\epsilon_1}, \qquad \beta \leq \ln \frac{1-\epsilon_2}{\epsilon_1}. \tag{3.4}$$

These relations are valid even if the observations are not independently distributed as long as SPRT terminates with probability 1.

Although (3.4) does not give exact values for α and β, these bounds can be used as approximate values for them. In fact, if $\mathcal{L}_k = \alpha$ or $\mathcal{L}_k = \beta$ when the test is terminated, then (3.4) will hold with equalities. In other words, the approximations

$$\alpha \approx \ln \frac{\epsilon_2}{1-\epsilon_1}, \qquad \beta \approx \ln \frac{1-\epsilon_2}{\epsilon_1} \tag{3.5}$$

neglect the overshoot of the log likelihood ratio beyond the thresholds. This overshoot is fairly small if the increments z_k are small in magnitude compared to the thresholds. In view of (3.3), this is the case if $\mathcal{H}_0$ and $\mathcal{H}_1$ are close to each other. It can be shown (Wald, 1947) that the sum of the probabilities of the two types of error does not increase, and the increase in the number of observations necessary to reach a decision is not appreciable, if the approximations (3.5) are used instead of the exact values. Hence, the thresholds can be taken as in (3.5) for all practical purposes.

The SPRT described above does not take any prior knowledge about the hypotheses into account. Quantitatively, this would correspond to assuming $p_0(\theta_0) = p_0(\theta_1) = 1/2$. A comparison between (2.37) and (3.1) reveals that the a priori probabilities can be used by initializing the log likelihood ratio to

$$\mathcal{L}_0 = \ln \frac{1-p_0(\theta_0)}{p_0(\theta_0)}. \tag{3.6}$$

One can also use $\mathcal{L}_0 = 0$ as the initial value if the thresholds

$$\alpha = \ln \frac{\epsilon_2 \, p_0(\theta_0)}{(1-p_0(\theta_0))(1-\epsilon_1)} \qquad \beta = \ln \frac{p_0(\theta_0)(1-\epsilon_2)}{(1-p_0(\theta_0))\epsilon_1}$$

are used.

3.2 Performance and Properties

Undoubtedly, among the quantities which describe the performance of a statistical test are its error probabilities. In a sequential test between two hypotheses, which will terminate eventually, the probability of making the correct or wrong decision can be given in terms of the probability of deciding in favour of $\mathcal{H}_0$, since the probability of accepting $\mathcal{H}_1$ will be one minus that value. The probability of accepting $\mathcal{H}_0$ is a function of the true parameter value and is denoted usually as the *operating characteristic* (OC) function.

On the other hand, a distinctive feature of a sequential test is that the sample number required for a terminal decision depends on the outcome of the observations and, hence, is a random variable. Since different sequential tests designed to achieve the same error probabilities may require different numbers of observations to reach a decision, another important quantity characterizing the performance of a sequential test is the *average sample number* (ASN).

Before giving the formulae for the ASN and OC functions for SPRT, we shall first introduce a *fundamental identity*, which is useful in deriving them.

3.2.1 Fundamental identity

Assume that the moment generating function $E\left\{e^{tz_k}\right\}$ of the increments z_k exists for all real t and is denoted by $m(t)$. Obviously, the equation

$$m(t) = 1 \qquad (3.7)$$

is satisfied for $t = 0$. It can be shown (Ghosh, 1970) that, assuming there exists some $0 < \delta < 1$ such that $\Pr\{z_k < \ln(1-\delta)\} > 0$ and $\Pr\{z_k > \ln(1+\delta)\} > 0$, then (3.7) has a unique solution $t_1 \neq 0$ if $E\{z_k\} \neq 0$ and no such nonzero solution exists if $E\{z_k\} = 0$. Moreover, t_1 and $E\{z_k\}$ have opposite signs.

Let us denote the sample number required for SPRT to reach a decision by n. That is, n is the smallest integer such that either $\mathcal{L}_n \leq \alpha$ or $\mathcal{L}_n \geq \beta$. Then, n is a random variable and the Fundamental Identity (also known as *Wald's identity*) is established by the following theorem.

Theorem 3.1 *If $\{z_k\}$ is a sequence of independently and identically distributed random variables, then*

$$E\left\{e^{t\mathcal{L}_n}[m(t)]^{-n}\right\} = 1 \tag{3.8}$$

for all $t \in \mathbb{R}$.

The proof of this result is originally due to Wald (1947) and is omitted here.

Note that the fundamental identity (3.8) holds trivially if n is fixed, i.e., in the case of a fixed sample size likelihood ratio test.

It can also be shown (Albert, 1947) that the left hand side of (3.8) can be differentiated with respect to t under the expectation operator any number of times.

3.2.2 ASN and OC functions

In SPRT, a decision is made in favour of $\mathcal{H}_0$ if $\mathcal{L}_k$ crosses the lower threshold α. Therefore, we shall denote the OC function of SPRT as $P_\alpha(\theta)$. Obviously, $P_\alpha(\theta_0) = 1 - \epsilon_1$ and $P_\alpha(\theta_1) = \epsilon_2$.

Equation (3.8) can be used to derive an expression for $P_\alpha(\theta)$. By putting $t_1(\theta)$ in (3.8), we get

$$E\left\{e^{t_1(\theta)\mathcal{L}_n}\right\} = 1$$

or

$$P_\alpha(\theta)E\{e^{t_1(\theta)\mathcal{L}_n} \mid \mathcal{L}_n \leq \alpha\} + (1 - P_\alpha(\theta))E\{e^{t_1(\theta)\mathcal{L}_n} \mid \mathcal{L}_n \geq \beta\} = 1. \tag{3.9}$$

If one neglects the overshoot of $\mathcal{L}_n$ beyond the thresholds and assumes that either $\mathcal{L}_n \approx \alpha$ or $\mathcal{L}_n \approx \beta$ when the test terminates, the OC function can be solved from (3.9) as

$$P_\alpha(\theta) \approx \frac{1 - e^{t_1(\theta)\beta}}{e^{t_1(\theta)\alpha} - e^{t_1(\theta)\beta}}. \tag{3.10}$$

If $\mathcal{H}_1$ is true, then we have from (3.3)

$$m(t) = \int_{-\infty}^{\infty} \left[\frac{f_1(y)}{f_0(y)}\right]^t f_1(y)\, dy.$$

Chapter 3. The Sequential Probability Ratio Test

Hence, (3.7) is satisfied for $t_1(\theta_1) = -1$. Similarly, $t_1(\theta_0) = 1$. Substituting these values into (3.10) we get the values of the OC function under the two hypotheses as

$$P_\alpha(\theta_0) \approx \frac{1 - e^\beta}{e^\alpha - e^\beta} \qquad P_\alpha(\theta_1) \approx \frac{1 - e^{-\beta}}{e^{-\alpha} - e^{-\beta}}. \tag{3.11}$$

Note that the equalities $P_\alpha(\theta_0) = 1 - \epsilon_1$ and $P_\alpha(\theta_1) = \epsilon_2$ can be confirmed by using the approximate values of α and β, given in (3.5), in (3.11).

The fundamental identity can also be used to derive the ASN formula. By differentiating both sides of (3.8) once with respect to t, we obtain

$$E\left\{\mathcal{L}_n e^{t\mathcal{L}_n}[m(t)]^{-n}\right\} - E\left\{e^{t\mathcal{L}_n} n[m(t)]^{-n-1} m'(t)\right\} = 0. \tag{3.12}$$

Evaluating (3.12) at $t = 0$ and noting that $m'(0) = E\{z_k\}$, one can solve for $E\{n\}$ as

$$E\{n\} = \frac{E\{\mathcal{L}_n\}}{E\{z_k\}} \tag{3.13}$$

when $E\{z_k\} \neq 0$. For the case where $E\{z_k\} = 0$, the ASN can be obtained by differentiating (3.8) twice and putting $t = 0$, which results in

$$E\{n\} = \frac{E\{\mathcal{L}_n^2\}}{E\{z_k^2\}}.$$

In many cases, it may not be possible to evaluate $E\{\mathcal{L}_n\}$ exactly. Nevertheless, (3.13) can be approximated as

$$E\{n\} \approx \frac{\alpha P_\alpha(\theta) + \beta(1 - P_\alpha(\theta))}{E\{z_k\}}.$$

Specifically, under $\mathcal{H}_0$ or $\mathcal{H}_1$

$$E\{n \mid \mathcal{H}_0\} \approx \frac{\alpha(1 - \epsilon_1) + \beta\epsilon_1}{E\{z_k \mid \mathcal{H}_0\}} \tag{3.14}$$

and

$$E\{n \mid \mathcal{H}_1\} \approx \frac{\alpha\epsilon_2 + \beta(1 - \epsilon_2)}{E\{z_k \mid \mathcal{H}_1\}} \tag{3.15}$$

will be the average sample numbers for SPRT.

3.3 Examples

In this section we apply SPRT to two examples.

Example 3.1 Consider the problem of deciding on the mean of a normally distributed random variable, y, having variance 1. So, the samples of y, taken sequentially, are independently distributed as

$$f(y(k)) = \frac{1}{\sqrt{2\pi}} \exp\left\{-\frac{1}{2}(y(k) - \theta)^2\right\}$$

and the hypotheses are described as $\mathcal{H}_0 : \theta = \theta_0$ and $\mathcal{H}_1 : \theta = \theta_1$ ($\theta_0 \neq \theta_1$).

The likelihood ratio is computed recursively by adding the scores

$$\begin{aligned} z_k &= \ln \frac{\frac{1}{\sqrt{2\pi}} \exp\left\{-\frac{1}{2}(y(k) - \theta_1)^2\right\}}{\frac{1}{\sqrt{2\pi}} \exp\left\{-\frac{1}{2}(y(k) - \theta_0)^2\right\}} \\ &= (\theta_1 - \theta_0)y(k) + \frac{1}{2}(\theta_0^2 - \theta_1^2) \end{aligned} \quad (3.16)$$

at each sampling instant; they are normally distributed, as is $y(k)$, and have the mean value

$$E\{z_k\} = (\theta_1 - \theta_0)\theta + \frac{1}{2}(\theta_0^2 - \theta_1^2). \quad (3.17)$$

Their moment generating function $m(t)$ can be given by

$$\begin{aligned} m(t) &= \int_{-\infty}^{\infty} \exp\left\{(\theta_1 - \theta_0)yt + \frac{1}{2}(\theta_0^2 - \theta_1^2)t\right\} \frac{1}{\sqrt{2\pi}} \exp\left\{-\frac{1}{2}(y - \theta)^2\right\} dy \\ &= \exp\left\{\frac{\theta_1 - \theta_0}{2}[(\theta_1 - \theta_0)t^2 - (\theta_1 + \theta_0 - 2\theta)t]\right\}. \end{aligned}$$

Hence, $m(t_1) = 1$ where

$$t_1 = \frac{\theta_1 + \theta_0 - 2\theta}{\theta_1 - \theta_0} \neq 0 \qquad \text{if} \quad \theta \neq (\theta_1 + \theta_0)/2.$$

Note that for $\theta = (\theta_1 + \theta_0)/2$, where t_1 turns out to be zero, we also have from (3.17), $E\{z_k\} = 0$.

Table 3.1 displays the results of Monte-Carlo simulations based on 10000 samples where the nominal values of ϵ_1, ϵ_2, $E\{n \mid \mathcal{H}_0\}$ and $E\{n \mid \mathcal{H}_1\}$ (i.e. the values calculated from the approximations (3.5), (3.14) and (3.15)) and their estimated

Chapter 3. The Sequential Probability Ratio Test   47

$\alpha = -2.197, \beta = 2.197$	ϵ_1	ϵ_2	$E\{n \mid \mathcal{H}_0\}$	$E\{n \mid \mathcal{H}_1\}$
Nominal	0.10	0.10	87.9	87.9
Estimated	0.09	0.07	92.0	91.7

$\alpha = -2.890, \beta = 2.251$	ϵ_1	ϵ_2	$E\{n \mid \mathcal{H}_0\}$	$E\{n \mid \mathcal{H}_1\}$
Nominal	0.10	0.05	118.2	99.7
Estimated	0.09	0.03	118.8	104.3

Table 3.1: Performance of SPRT in testing the mean of a normal distribution

values can be compared. The hypotheses are chosen as $\theta_0 = 0.0$ and $\theta_1 = 0.2$. Note that the estimated error probabilities are less than the values obtained from (3.5) and the cost for this is a slight increase in the ASN.

Example 3.2 As a second case, we consider the sequential decision problem where the probability distributions of y under $\mathcal{H}_0$ and $\mathcal{H}_1$ are given as in (2.38) and (2.39), respectively, and assume that $u(k) = u_1 \; \forall k$, which was proved to be the optimal input for the problem in Section 2.5. For simplicity, let us denote $\bar{q} = q + u_1$ and $\bar{r} = r + u_1$. In view of (2.45), we have $\bar{q} = 0.65$ and $\bar{r} = 0.85$.

It was shown in Section 2.5 that if the thresholds

$$\alpha = -0.543 \quad \text{and} \quad \beta = 0.916 \tag{3.18}$$

are used then SPRT will terminate after at most 4 samples are taken. First let us find the error probabilities when these thresholds are used. Certainly, one cannot assume that ϵ_1 and ϵ_2 associated with these thresholds are approximated using (3.5), since (2.52) reveals that $\mathcal{L}_k$ attains infinite values when terminating the test if $y(k) \neq 1$, let alone being close to the thresholds. In fact, $\epsilon_2 = 0$. Because, when $\mathcal{H}_1$ is true, $y(k)$ can be either 1 or 2, hence, $\mathcal{L}_k$ can take only positive values (see (2.52)) and cannot reach the negative threshold α. On the other hand, when $\mathcal{H}_0$ is true, $\mathcal{H}_1$ is accepted only if $y(i) = 1$, $i = 1, \ldots, 4$. Hence $\epsilon_1 = \bar{q}^4 = 0.179$.

Secondly, we shall compute the ASN. The formula (3.13) is not directly applicable in this case, since in view of (2.52), $E\{\mathcal{L}_n\}$ is not finite under either hypothesis.

Neither is $E\{z_k\}$, since

$$z_k = \begin{cases} -\infty & \text{if } y(k) = 0 \\ \ln(\bar{q}/\bar{r}) & \text{if } y(k) = 1 \\ \infty & \text{if } y(k) = 2. \end{cases}$$

Therefore, we shall consider the following probability density functions:

$$f_{\varepsilon 0}(y(k)) = \begin{cases} 1 - \bar{q} - \varepsilon & \text{if } y(k) = 0 \\ \bar{q} & \text{if } y(k) = 1 \\ \varepsilon & \text{if } y(k) = 2 \end{cases} \quad (3.19)$$

and

$$f_{\varepsilon 1}(y(k)) = \begin{cases} \varepsilon & \text{if } y(k) = 0 \\ \bar{r} & \text{if } y(k) = 1 \\ 1 - \bar{r} - \varepsilon & \text{if } y(k) = 2 \end{cases} \quad (3.20)$$

instead of f_0 and f_1. Here, $0 < \varepsilon < \delta < 1$ where

$$\ln \frac{\delta}{1 - \bar{q} - \delta} < \alpha \quad \text{and} \quad \ln \frac{1 - \bar{r} - \delta}{\delta} > \beta. \quad (3.21)$$

Clearly, (3.19) and (3.20) tend to (2.38) and (2.39) as $\varepsilon \to 0$. The conditions in (3.21) guarantee that SPRT is terminated whenever $y(k) \neq 1$. Hence, a decision will still be reached in at most 4 sampling instants if the thresholds are as in (3.18). With these modified probabilities the mean value of z_k under $\mathcal{H}_0$ is found to be

$$E\{z_k \mid \mathcal{H}_0\} = \bar{q} \ln \frac{\bar{r}}{\bar{q}} + g(\varepsilon) \quad (3.22)$$

with

$$g(\varepsilon) = (1 - \bar{q} - \varepsilon) \ln \frac{\varepsilon}{1 - \bar{q} - \varepsilon} + \varepsilon \ln \frac{1 - \bar{r} - \varepsilon}{\varepsilon}.$$

The test terminates at $n_\varepsilon < 4$ if one observes $y(n_\varepsilon) \neq 1$ following $y(i) = 1$ for $i < n_\varepsilon$. If this is not the case, it is terminated at $n_\varepsilon = 4$. Hence, the mean of the log likelihood ratio at the end of the test is computed as

$$E\{\mathcal{L}_{n_\varepsilon} \mid \mathcal{H}_0\}$$
$$= \sum_{i=1}^{4} \Pr\{n_\varepsilon = i\} \mathcal{L}_i$$
$$= \sum_{i=0}^{3} \bar{q}^i \left[(1 - \bar{q} - \varepsilon) \ln \frac{\bar{r}^i \varepsilon}{\bar{q}^i (1 - \bar{q} - \varepsilon)} + \varepsilon \ln \frac{\bar{r}^i (1 - \bar{r} - \varepsilon)}{\bar{q}^i \varepsilon} \right] + \bar{q}^4 \ln \frac{\bar{q}^4}{\bar{r}^4}.$$

Rearranging, we get

$$E\{\mathcal{L}_{n_\varepsilon} \mid \mathcal{H}_0\} = g(\varepsilon)\frac{1-\bar{q}^4}{1-\bar{q}} + (1-\bar{q})\ln\frac{\bar{r}}{\bar{q}} \sum_{i=0}^{3} i\bar{q}^i + 4\bar{q}^4 \ln\frac{\bar{r}}{\bar{q}}. \qquad (3.23)$$

The ASN (under $\mathcal{H}_0$) of the SPRT between f_0 and f_1 can be found by

$$E\{n \mid \mathcal{H}_0\} = \lim_{\varepsilon \to 0} \frac{E\{\mathcal{L}_{n_\varepsilon}\}}{E\{z_k\}}. \qquad (3.24)$$

Substituting (3.23) and (3.22) into (3.24) and noting that $\lim_{\varepsilon \to 0} g(\varepsilon) = -\infty$, we get

$$E\{n \mid \mathcal{H}_0\} = \frac{1-\bar{q}^4}{1-\bar{q}} = 2.35. \qquad (3.25)$$

A similar application of this limiting approach under $\mathcal{H}_1$ gives

$$E\{n \mid \mathcal{H}_1\} = \frac{1-\bar{r}^4}{1-\bar{r}} = 3.19. \qquad (3.26)$$

Note that no approximation is involved in (3.25) and (3.26).

These results are confirmed on a simulation with 1000 runs, which yielded the estimated values $E\{n \mid \mathcal{H}_0\} = 2.32$ and $E\{n \mid \mathcal{H}_1\} = 3.17$.

3.4 Conclusions

In this chapter we gave a brief overview of some essential results about the SPRT in the case of i.i.d. observations. In Section 1, one of the appealing properties of SPRT is given; namely that the test thresholds are obtained using the error probabilities only, and are independent of the nature of the hypotheses. We also introduced a fundamental identity which has been useful in deriving the ASN and the OC function.

Let us emphasize that most of the quantities associated with SPRT, such as α, β, ASN and OC, can be obtained, in general, only approximately and the basic assumption underlying these approximations is that the overshoot of the log likelihood ratio beyond the thresholds, when the test is finished, is negligible. Example 3.2 illustrates a case where the log likelihood ratio takes values from a finite set, making it possible to obtain the exact value of its mean at termination. This, in turn, can be used to evaluate the ASN exactly.

In this chapter, we have restricted ourselves to the i.i.d. case (except that (3.5) is valid generally). However, this case is not relevant if the hypotheses describe the data generators as dynamic (correlated) processes. Therefore, our framework must be extended to include the non-i.i.d. case. This goal will be pursued in the next chapter.

Chapter 4

Sequential Analysis of Autoregressive Processes

One of the features of SPRT peculiar to the case when deciding about the parameters describing the *dynamics* of a process, is that the observations are no longer i.i.d.. This chapter investigates the performance of SPRT in such a case, namely, the autoregressive one.

We define our models and hypotheses in Section 1. Section 2 presents some lemmas which will be used in deriving an analogue of the Fundamental Identity, the ASN and OC function. These include propositions about the asymptotic behaviour of the log likelihood ratio and the introduction of a function $\lambda(t)$ which is basically an analogue of the moment generating function $m(t)$. The derivation of the Fundamental Identity corresponding to the autoregressive (AR) case is given in Section 3, and the ASN and OC function, in Section 4. All the analysis in Sections 2-4 is based on the likelihood ratio conditioned on the initial values of the observations. Section 5 consists of some remarks indicating that the same analysis can be carried out by employing exact likelihood functions. After two simulation examples in Section 6, some conclusions are drawn in Section 7.

4.1 Introduction

In this chapter, we assume that the observations are generated by the stationary autoregressive (AR) model

$$A(q^{-1})y(k) = \epsilon(k) \tag{4.1}$$

where

$$A(q^{-1}) = 1 + a_1 q^{-1} + \cdots + a_{n_a} q^{-n_a} \tag{4.2}$$

is a polynomial in the backward shift operator q^{-1} with a known degree n_a and $A(z)$ has all its zeros outside the unit circle. Here, $\epsilon(k)$ is a white Gaussian random noise with zero mean and variance σ^2.

The sequential probability ratio test can be applied to decide between two hypotheses about the parameter vector $\boldsymbol{\theta} = [a_1, \ldots, a_{n_a}]^T$;

$$\mathcal{H}_0: \quad \boldsymbol{\theta} = \boldsymbol{\theta}_0 \triangleq [a_{01}, \ldots, a_{0n_a}]^T, \tag{4.3}$$

$$\mathcal{H}_1: \quad \boldsymbol{\theta} = \boldsymbol{\theta}_1 \triangleq [a_{11}, \ldots, a_{1n_a}]^T. \tag{4.4}$$

The recursive computation of the log likelihood ratio in (3.2) is still valid, provided z_k is defined by

$$z_k = \begin{cases} \ln \dfrac{f_1(y(k) \mid y(k-1), \ldots, y(k-n_a))}{f_0(y(k) \mid y(k-1), \ldots, y(k-n_a))} & \text{for} \quad k \geq n_a + 1 \\ \ln \dfrac{f_1(y(k) \mid y(k-1), \ldots, y(1))}{f_0(y(k) \mid y(k-1), \ldots, y(1))} & \text{for} \quad n_a \geq k \geq 2 \\ \ln \dfrac{f_1(y(1))}{f_0(y(1))} & \text{for} \quad k = 1 \end{cases} \tag{4.5}$$

and the test is conducted by monitoring the inequality (3.1).

If the hypotheses are close to each other, so that z_k is small in magnitude (hence, a large sample number is required to reach a decision), the first n_a scores can either be ignored or computed as those for $k > n_a$. So, the increments can be calculated in practice by

$$z_k = \ln \dfrac{f_1(y(k) \mid y(k-1), \ldots, y(k-n_a))}{f_0(y(k) \mid y(k-1), \ldots, y(k-n_a))} \quad \text{for} \quad k \geq 1. \tag{4.6}$$

Chapter 4. Sequential Analysis of AR Processes

To start the test one can use an a priori estimate for $\mathbf{y}_0 = [y(1-n_a), \ldots, y(0)]^T$, or one would have observed the initial values already. In this case the use of recursion in (3.2) yields the *conditional log likelihood ratio*

$$\mathcal{L}_k = \ln \frac{f_1(\mathbf{y}_k \mid \mathbf{y}_0)}{f_0(\mathbf{y}_k \mid \mathbf{y}_0)}$$

at each sampling instant.

It is seen from (4.5) or (4.6) that the increments depend on the past data and hence, in general, are not independent. Therefore, the results presented in the previous chapter about the properties of the SPRT should be extended to include autoregressive models. The ASN and OC function can be obtained if one has an analogue of the Fundamental Identity at hand.

The extension of Wald's results and his Fundamental Identity to several specific non-i.i.d. cases has been considered by many authors. Tweedie (1960) and Miller (1962) considered an extension to the case where the observations are obtained from a finite Markov chain. Similar results for the same case have been found by Phatarfod (1965). Later, Phatarfod (1971) derived the Fundamental Identity also for the Gauss-Markov case. In fact, most of these works are inspired by an earlier note of Bellman (1957), where possible generalizations of Wald's identity are hinted at.

As far as the general non-i.i.d. case is concerned, Ghosh (1970) conjectured that the expressions for the ASN and OC function are of the same form as those for the i.i.d. case (i.e., (3.10) and (3.13)). This conjecture can be justified by the results of Eisenberg and Ghosh (1979) where the Fundamental Identity and various sampling properties of sequential tests are derived for a quite general case. Eisenberg and Ghosh (1979) showed that various identities and approximations in sequential analysis follow directly from the definition of the likelihood ratio, without making any reference to the dependency of observations or even to the exact structure of the sequential test.

To be more constructive here and hence to clarify the relations between different quantities associated with SPRT, we present below an alternative derivation of the Fundamental Identity, rather than directly specializing the results of Eisenberg and Ghosh (1979) to the AR case. We will extend the framework introduced in Chapter 3 by exploiting the correlation structure of the observations.

4.2 Some Lemmas

4.2.1 Asymptotic behaviour of conditional log likelihood ratio

The results obtained for discrete Markov chains and Gauss-Markov processes by Phatarfod (1965; 1971) suggest that the key point in deriving an analogue of the Fundamental Identity is to show that $M_k(t)$, the moment generating function of $\mathcal{L}_k$, can be written as $M_k(t) \sim C(t) \lambda^k(t)$ for large k where $C(t)$ is independent of k.

To derive the moment generating function of the likelihood ratio we will first find the conditional distribution of the observation vector $\mathbf{y}_k$ following Box and Jenkins (1976; Ch. 7) (See also Pesaran and Slater, 1980). By using (4.1), we can write

$$\mathbf{e}_k = \mathbf{Q}\mathbf{y}_a$$

where $\mathbf{e}_k = [\epsilon(1), \ldots, \epsilon(k)]^T$, $\mathbf{y}_a^T = [\mathbf{y}_0^T \ \mathbf{y}_k^T]$ and $\mathbf{Q}$ is an $k \times (k+n_a)$ matrix given by

$$\mathbf{Q} = \begin{bmatrix} a_{n_a} & a_{n_a-1} & \cdots & \cdots & 1 & 0 & & 0 \\ 0 & a_{n_a} & \cdots & \cdots & a_1 & 1 & & \\ & & \ddots & & & & \ddots & \\ 0 & & & & a_{n_a} & & & 1 \end{bmatrix}. \quad (4.7)$$

Since the Jacobian of the transformation $\mathbf{e}_k \mapsto \mathbf{y}_k$ is unity the conditional joint distribution of $\mathbf{y}_k$ is

$$f(\mathbf{y}_k \mid \mathbf{y}_0) = \frac{1}{(\sigma\sqrt{2\pi})^k} \exp\left\{-\frac{1}{2\sigma^2} \mathbf{y}_a^T \mathbf{Q}^T \mathbf{Q} \mathbf{y}_a\right\}. \quad (4.8)$$

Assume that the initial values $\{y(-n_a+1), \ldots, y(0)\}$ have the stationary distribution of n_a consecutive samples from (4.1), i.e.,

$$f(\mathbf{y}_0) = \frac{|\mathbf{D}|^{1/2}}{(\sigma\sqrt{2\pi})^{n_a}} \exp\left\{-\frac{1}{2\sigma^2} \mathbf{y}_0^T \mathbf{D} \mathbf{y}_0\right\}. \quad (4.9)$$

with $\sigma^2 \mathbf{D}^{-1}$ being the $n_a \times n_a$ covariance matrix of $\mathbf{y}_0$. Then by using (4.8) and (4.9) the probability distribution of $\mathbf{y}_a$ can be written as

$$f(\mathbf{y}_a) = f(\mathbf{y}_k \mid \mathbf{y}_0) f(\mathbf{y}_0)$$
$$= \frac{|\mathbf{D}|^{1/2}}{(\sigma\sqrt{2\pi})^{k+n_a}} \exp\left\{-\frac{1}{2\sigma^2} \mathbf{y}_a^T \mathbf{V} \mathbf{y}_a\right\} \qquad (4.10)$$

where

$$\mathbf{V} = \begin{bmatrix} \mathbf{D} & 0 \\ 0 & 0 \end{bmatrix} + \mathbf{Q}^T \mathbf{Q}. \qquad (4.11)$$

is a $(k+n_a) \times (k+n_a)$ matrix.

Note that $\mathbf{V}/\sigma^2$ is the inverse of the covariance matrix of the random vector $\mathbf{y}_a$, and hence is doubly symmetric, (i.e., symmetric with respect to both of its diagonals) (Box and Jenkins, 1976). Such a matrix satisfies $\mathbf{J}_{k+n_a} \mathbf{V} \mathbf{J}_{k+n_a} = \mathbf{V}$ where $\mathbf{J}_{k+n_a}$ is the $(k+n_a) \times (k+n_a)$ matrix with 1's on the secondary diagonal and 0's elsewhere (Roebuck and Barnett, 1978). Premultiplying a matrix by $\mathbf{J}$ reverses the order of its rows and postmultiplying it by $\mathbf{J}$ reverses the order of its columns. In view of (4.11), the double symmetry of $\mathbf{V}$ can be used to determine the elements of $\mathbf{D}$ from the lower right hand $n_a \times n_a$ corner of $\mathbf{Q}^T \mathbf{Q}$. Further, it follows from (4.11) and (4.7) that $\mathbf{V}$ is banddiagonal with a bandwidth of $2n_a + 1$. Another property of $\mathbf{V}$ is that

$$[\mathbf{V}]_{ij} = [\mathbf{V}]_{i-j} \quad \text{if} \quad \max(i,j) > n_a \quad \text{and} \quad \min(i,j) < k - n_a + 1. \qquad (4.12)$$

That means, $\mathbf{V}$ is of Toeplitz type except at its $n_a \times n_a$ upper left and lower right hand corners. In the sequel, we use the term *nearly Toeplitz* exclusively for matrices satisfying (4.12). We will refer to the Toeplitz matrix obtained from a nearly Toeplitz one by changing its defective corners as its *associated Toeplitz matrix*. Note that a symmetric Toeplitz matrix is always doubly symmetric.

Let us denote by $\mathbf{Q}_i$ the value of $\mathbf{Q}$ under the hypothesis $\mathcal{H}_i$ (i=0,1). Then, from (4.8) and (4.10), the moment generating function of the conditional log likelihood ratio is

$$M_k(t) = E\left\{e^{\mathcal{L}_k t}\right\}$$

$$= E\left\{\left[\frac{f_1(\mathbf{y}_k \mid \mathbf{y}_0)}{f_0(\mathbf{y}_k \mid \mathbf{y}_0)}\right]^t\right\}$$

$$= \frac{|\mathbf{D}|^{1/2}}{(\sigma\sqrt{2\pi})^{k+n_a}} \int_{-\infty}^{\infty} \exp\left\{-\frac{1}{2\sigma^2}\mathbf{y}_a^T[(\mathbf{Q}_1^T\mathbf{Q}_1 - \mathbf{Q}_0^T\mathbf{Q}_0)t + \mathbf{V}]\mathbf{y}_a\right\} d\mathbf{y}_a. \tag{4.13}$$

For values of t such that $(\mathbf{Q}_1^T\mathbf{Q}_1 - \mathbf{Q}_0^T\mathbf{Q}_0)t + \mathbf{V}$ is positive definite

$$M_k(t) = \frac{|\mathbf{D}|^{1/2}}{|(\mathbf{Q}_1^T\mathbf{Q}_1 - \mathbf{Q}_0^T\mathbf{Q}_0)t + \mathbf{V}|^{1/2}}. \tag{4.14}$$

Note that, since $\mathbf{V}$ is positive definite, the values of t for which (4.14) is valid include a nonempty interval around 0. We denote this interval by Υ.

Equation (4.14) reveals that in order to analyze the behaviour of $M_k(t)$ for large k we have to consider the asymptotic behaviour of the determinants of nearly Toeplitz matrices. First we will quote a famous result on the asymptotic behaviour of Toeplitz determinants due to Grenander and Szegö (1958).

Theorem 4.1 *Let $\mathbf{T}_k = [c_{i-j}]_{i,j=1}^k$ be a symmetric positive definite Toeplitz matrix $(k = 1, 2, \ldots)$. Let $g(z)$ be a polynomial such that*

$$|g(e^{j\omega})|^2 = c_0 + 2\sum_{n=1}^{\infty} c_n \cos n\omega \tag{4.15}$$

and $g(z)$ has no zeros inside the unit circle. Then,

$$\lim_{k\to\infty} \frac{|\mathbf{T}_k|}{\mathcal{G}^k} = \Omega$$

where $\mathcal{G}$ is the geometric mean of $|g(e^{j\omega})|^2$, i.e.,

$$\mathcal{G} = \exp\left\{\frac{1}{2\pi}\int_{-\pi}^{\pi} \ln|g(e^{j\omega})|^2 d\omega\right\} \tag{4.16}$$

and

$$\Omega = \exp\left\{\frac{1}{\pi}\iint_{|z|\leq 1} \left|\frac{g'(z)}{g(z)}\right|^2 ds\right\} \tag{4.17}$$

with ds being area element on the complex plane.

Chapter 4. Sequential Analysis of AR Processes

The proof can be found in Grenander and Szegö (1958). In fact, it applies to a more general class of Toeplitz matrices than stated here.

Note that in the case of banddiagonal Toeplitz matrices $g(z)$ will be a finite polynomial.

Next we present a lemma on nearly Toeplitz matrices which will be useful in deriving the Fundamental Identity.

Lemma 4.1 *Let $\{\mathbf{A}_N\}_{N=2n_a+1}^{\infty}$ be a sequence of $N \times N$ banddiagonal (with a bandwidth of $2n_a + 1$), symmetric, nearly Toeplitz matrices with identical $(n_a + 1)$-st leading submatrices[1] and $\{\mathbf{T}_N\} = \{[c_{i-j}]_{ij=1}^{N}\}$ be the associated Toeplitz matrices. Further let $\mathcal{G}$ and Ω be defined as in (4.16) and (4.17). Then*

$$|\mathbf{A}_N| \sim |\mathbf{G}_1 - \mathbf{LS}_{n_a\infty}\mathbf{L}^T||\mathbf{G}_2 - \mathbf{LS}_{n_a\infty}\mathbf{L}^T|\,\Omega\,\mathcal{G}^{N-2n_a} \qquad \text{for large } N$$

where $\mathbf{G}_1$ and $\mathbf{G}_2$ are the n_a-th leading submatrices of $\mathbf{A}_N$ and $\mathbf{J}_N\mathbf{A}_N\mathbf{J}_N$, respectively,

$$\mathbf{L} = \begin{bmatrix} c_{n_a} & & 0 \\ \vdots & \ddots & \\ c_1 & \cdots & c_{n_a} \end{bmatrix}$$

and $\mathbf{S}_{n_a\infty} = \lim_{N\to\infty}\mathbf{S}_{n_aN}$; $\mathbf{S}_{iN}$ being the i-th leading submatrix of $\mathbf{T}_N^{-1}$.

Proof. By using its nearly Toeplitz character, $\mathbf{A}_N$ can be partitioned for $N > 3n_a$ as

$$\mathbf{A}_N = \begin{bmatrix} \mathbf{G}_1 & \mathbf{X}^T & \mathbf{0}_{n_a \times n_a} \\ \mathbf{X} & \mathbf{T}_{N-2n_a} & \mathbf{J}_{N-2n_a}\mathbf{X}\mathbf{J}_{n_a} \\ \mathbf{0}_{n_a \times n_a} & \mathbf{J}_{n_a}\mathbf{X}^T\mathbf{J}_{N-2n_a} & \mathbf{J}_{n_a}\mathbf{G}_2\mathbf{J}_{n_a} \end{bmatrix}$$

where

$$\mathbf{X}^T = [\mathbf{L} \ \mathbf{0}_{n_a \times N-3n_a}] \qquad (4.18)$$

and $n \times l$ zero matrices are denoted by $\mathbf{0}_{n \times l}$. By denoting

$$\mathbf{B}_{N-n_a} = \begin{bmatrix} \mathbf{T}_{N-2n_a} & \mathbf{J}_{N-2n_a}\mathbf{X}\mathbf{J}_{n_a} \\ \mathbf{J}_{n_a}\mathbf{X}^T\mathbf{J}_{N-2n_a} & \mathbf{J}_{n_a}\mathbf{G}_2\mathbf{J}_{n_a} \end{bmatrix},$$

---
[1] By the n-th leading submatrix of $\mathbf{A} = [a_{ij}]_{i,j=1}^{k}$ we mean the $n \times n$ matrix $[a_{ij}]_{i,j=1}^{n}$.

and using the rules for determinants of partitioned matrices we can write

$$|\mathbf{A}_N| = \left|\mathbf{G}_1 - [\mathbf{X}^T \ \mathbf{0}_{n_a \times n_a}]\mathbf{B}_{N-n_a}^{-1}\begin{bmatrix}\mathbf{X}\\ \mathbf{0}_{n_a \times n_a}\end{bmatrix}\right|$$

$$\times \left|\mathbf{J}_{n_a}\mathbf{G}_2\mathbf{J}_{n_a} - \mathbf{J}_{n_a}\mathbf{X}^T\mathbf{J}_{N-2n_a}\mathbf{T}_{N-2n_a}^{-1}\mathbf{J}_{N-2n_a}\mathbf{X}\mathbf{J}_{n_a}\right| |\mathbf{T}_{N-2n_a}|.$$

Since $\mathbf{T}^{-1}$ has a double symmetry inherited from $\mathbf{T}$, it follows that $\mathbf{J}\mathbf{T}^{-1}\mathbf{J} = \mathbf{T}^{-1}$. Hence, by using the definition of $\mathbf{X}$ in (4.18) and $|\mathbf{J}| = 1$, we obtain

$$|\mathbf{A}_N| = |\mathbf{G}_1 - \mathbf{L}\mathbf{Z}_{n_a,N-n_a}\mathbf{L}^T||\mathbf{G}_2 - \mathbf{L}\mathbf{S}_{n_a,N-n_a}\mathbf{L}^T||\mathbf{T}_{N-2n_a}| \qquad (4.19)$$

where $\mathbf{Z}_{i,N-n_a}$ stands for the i-th leading submatrix of $\mathbf{B}_{N-n_a}^{-1}$. In fact, since $\mathbf{B}$ and $\mathbf{T}$ differ from each other only in their lower right hand $n_a \times n_a$ corners we will have the elementwise limit

$$\mathbf{Z}_{n_a N} \to \mathbf{S}_{n_a N} \quad \text{as} \quad N \to \infty. \qquad (4.20)$$

On the other hand, from Theorem 4.1 we have

$$|\mathbf{T}_{N-2n_a}| \sim \Omega \mathcal{G}^{N-2n_a} \quad \text{for large } N. \qquad (4.21)$$

So, using (4.20) and (4.21) in (4.19), we obtain

$$|\mathbf{A}_N| \sim |\mathbf{G}_1 - \mathbf{L}\mathbf{S}_{n_a\infty}\mathbf{L}^T||\mathbf{G}_2 - \mathbf{L}\mathbf{S}_{n_a\infty}\mathbf{L}^T|\Omega \mathcal{G}^{N-2n_a} \quad \text{for large } N.$$

$\square$

Considering $(\mathbf{Q}_1^T\mathbf{Q}_1 - \mathbf{Q}_0^T\mathbf{Q}_0)t + \mathbf{V}$ as the matrix $\mathbf{A}_{k+n_a}$ in this lemma and using (4.14) yields the following asymptotic expression for $M_k(t)$:

$$M_k(t) \sim C(t)\,\lambda^k(t) \qquad (4.22)$$

where

$$C(t) = \left[\frac{|\mathbf{D}|\,\mathcal{G}^{n_a}(t)}{|\mathbf{G}_1(t) - \mathbf{L}(t)\,\mathbf{S}_{n_a\infty}(t)\,\mathbf{L}^T(t)|\,|\mathbf{G}_2(t) - \mathbf{L}(t)\,\mathbf{S}_{n_a\infty}(t)\,\mathbf{L}^T(t)|\,\Omega(t)}\right]^{1/2} \qquad (4.23)$$

and

$$\lambda(t) = \mathcal{G}^{-1/2}(t). \qquad (4.24)$$

Let us illustrate this asymptotic formulation of $M_k(t)$ by two simple examples.

Example 4.1 First, we consider the case $n_a = 1$, i.e.,

$$y(k) + a_1 y(k-1) = \epsilon(k). \qquad (4.25)$$

From (4.7), we find

$$\mathbf{Q}^T \mathbf{Q} = \begin{bmatrix} a_1^2 & a_1 & & & \\ a_1 & 1+a_1^2 & \ddots & & \\ & \ddots & \ddots & \ddots & \\ & & \ddots & 1+a_1^2 & a_1 \\ & & & a_1 & 1 \end{bmatrix}.$$

By using the double symmetry of $\mathbf{V}$ in (4.11),

$$\mathbf{D} = 1 - a_1^2. \qquad (4.26)$$

So, the matrix $(\mathbf{Q}_1^T \mathbf{Q}_1 - \mathbf{Q}_0^T \mathbf{Q}_0)t + \mathbf{V}$ can be found as a tridiagonal nearly Toeplitz matrix

$$(\mathbf{Q}_1^T \mathbf{Q}_1 - \mathbf{Q}_0^T \mathbf{Q}_0)t + \mathbf{V} = \begin{bmatrix} 1+(a_{11}^2 - a_{01}^2)t & c_1(t) & & & \\ c_1(t) & c_0(t) & \ddots & & \\ & \ddots & \ddots & \ddots & \\ & & \ddots & c_0(t) & c_1(t) \\ & & & c_1(t) & 1 \end{bmatrix} \qquad (4.27)$$

where

$$c_0(t) = 1 + a_1^2 + (a_{11}^2 - a_{01}^2)t \qquad c_1(t) = (a_{11} - a_{01})t + a_1. \qquad (4.28)$$

By using (4.15), namely $|g(e^{j\omega}, t)|^2 = c_0(t) + 2c_1(t)\cos\omega$, or with $z = e^{j\omega}$

$$\begin{aligned} g(z,t)\,g(z^{-1},t) &= c_0(t) + c_1(t)z + c_1(t)z^{-1} \\ &= \mu_2(t)\left(z + \frac{\mu_1(t)}{c_1(t)}\right)\left(z^{-1} + \frac{\mu_1(t)}{c_1(t)}\right) \end{aligned}$$

where

$$\mu_1(t) = \frac{1}{2}\left[c_0(t) + \sqrt{c_0^2(t) - 4c_1^2(t)}\right] \quad \text{and} \quad \mu_2(t) = \frac{1}{2}\left[c_0(t) - \sqrt{c_0^2(t) - 4c_1^2(t)}\right], \qquad (4.29)$$

the polynomial $g(z,t)$ turns out to be

$$g(z,t) = \sqrt{\mu_2(t)} \left(z + \frac{\mu_1(t)}{c_1(t)} \right). \tag{4.30}$$

Substituting (4.30) in (4.16),

$$\mathcal{G}(t) = \exp\left\{ \frac{1}{2\pi} \int_{-\pi}^{\pi} \ln[c_0(t) + 2c_1(t)\cos\omega]\, d\omega \right\}.$$

Since

$$\int_0^\pi \ln(a + b\cos\omega)\, d\omega = \pi \ln\frac{a + \sqrt{a^2 - b^2}}{2} \qquad \text{for} \quad |a| > |b| \tag{4.31}$$

(Prudnikov et al., 1988), we obtain

$$\mathcal{G}(t) = \exp\left\{ \ln\frac{c_0(t) + \sqrt{c_0(t)^2 - 4c_1(t)^2}}{2} \right\} = \mu_1(t). \tag{4.32}$$

Also, by making use of (4.17),

$$\Omega(t) = \exp\left\{ \frac{1}{\pi} \iint_{|z|\leq 1} \left| \frac{1}{z + \frac{\mu_1(t)}{c_1(t)}} \right|^2 ds \right\}. \tag{4.33}$$

The area element of the integration on the complex plane is $ds = r\,dr\,d\omega$ in polar coordinates or by making the substitution $\zeta = e^{j\omega}$, $ds = r\,d\zeta\,dr/(j\zeta)$. Hence, the double integral in (4.33) can be written as

$$\iint_{|z|\leq 1} \left| \frac{1}{z + \frac{\mu_1(t)}{c_1(t)}} \right|^2 ds = \frac{1}{j} \int_0^1 \oint_{|\zeta|=1} \frac{r}{(r\zeta + x)(r\zeta^{-1} + x)\zeta} d\zeta\, dr$$

$$= \frac{1}{jx} \int_0^1 \oint_{|\zeta|=1} \frac{1}{(\zeta + x/r)(\zeta + r/x)} d\zeta\, dr \tag{4.34}$$

where

$$x = \sqrt{\frac{\mu_1(t)}{\mu_2(t)}}. \tag{4.35}$$

From (4.29), it follows that $|x| > 1$. Hence, inside the unit circle we have $|x/r| > 1$. So, by performing the integration with respect to ζ in (4.34) and rearranging, one gets

$$\Omega(t) = \exp\left\{ 2\int_0^1 \frac{r}{\mu_1(t)/\mu_2(t) - r^2} dr \right\}$$

$$= \frac{\mu_1(t)}{\mu_1(t) - \mu_2(t)}. \tag{4.36}$$

Chapter 4. Sequential Analysis of AR Processes 61

As is evident from (4.27), we have for this special case

$$\mathbf{G}_1(t) = 1 + (a_{11}^2 + a_{01}^2)t, \qquad \mathbf{G}_2(t) = 1, \qquad \mathbf{L}(t) = c_1(t). \tag{4.37}$$

Finally, $\mathbf{S}_{1\infty}(t)$ can be computed using the inversion formula for tridiagonal symmetric Toeplitz matrices given by Mentz (1976), namely

$$\mathbf{S}_{1k}(t) = \frac{x^{-2k} - 1}{x\, c_1(t)\, (x^{-2k-2} - 1)} \tag{4.38}$$

where x is given by (4.35). Hence, by taking the limit as $k \to \infty$ in (4.38) and noting that $c_1^2(t) = \mu_1(t)\mu_2(t)$,

$$\mathbf{S}_{1\infty}(t) = \frac{1}{xc_1(t)} = \frac{1}{\mu_1(t)}. \tag{4.39}$$

So, using (4.26), (4.32), (4.36) (4.37) and (4.39) in (4.23) and (4.24) we find

$$M_k(t) \sim \left[\frac{(1 - a_1^2)(\mu_1(t) - \mu_2(t))}{(1 + (a_{11}^2 - a_{01}^2)t - \mu_2(t))(1 - \mu_2(t))} \right]^{1/2} \mu_1^{-k/2}(t).$$

Example 4.2 Now, let us consider the second order autoregressive process

$$y(k) + a_2 y(k-2) = \epsilon(k).$$

In this case

$$(\mathbf{Q}_1^T\mathbf{Q}_1 - \mathbf{Q}_0^T\mathbf{Q}_0)t + \mathbf{V} = \begin{bmatrix} \mathbf{G}_1(t) & \begin{matrix} c_2(t) \\ 0 \end{matrix} & \ddots & & & \\ c_2(t) & 0 & c_0(t) & \ddots & \ddots & \\ & \ddots & \ddots & \ddots & \ddots & \\ & & \ddots & \ddots & c_0(t) & 0 & c_2(t) \\ & & & \ddots & 0 & 1 & 0 \\ & & & & c_2(t) & 0 & 1 \end{bmatrix} \tag{4.40}$$

with $\mathbf{G}_1(t) = (1 + (a_{12}^2 - a_{02}^2)t)\mathbf{I}_2$, $\mathbf{I}_2$ being the 2×2 identity matrix and

$$c_0(t) = 1 + a_2^2 + (a_{12}^2 - a_{02}^2)t \qquad c_2(t) = (a_{12} - a_{02})t + a_2. \tag{4.41}$$

The polynomial $g(z,t)$ can be determined from

$$g(z,t)g(z^{-1},t) = c_0(t) + c_2(t)z^2 + c_2(t)z^{-2}$$

Chapter 4. Sequential Analysis of AR Processes

and turns out to be

$$g(z,t) = \sqrt{\bar{\mu}_2(t)} \left(z^2 + \frac{\bar{\mu}_1(t)}{c_2(t)} \right)$$

where $\bar{\mu}_1(t)$ and $\bar{\mu}_2(t)$ are given as in (4.29) with $c_1(t)$ replaced by $c_2(t)$, and both $c_0(t)$ and $c_2(t)$ are given by (4.41). By performing the integrals in (4.16) and (4.17), we get $\mathcal{G}(t) = \sqrt{\bar{\mu}_1(t)}$ and $\Omega(t) = \bar{\mu}_1^2(t)/(\bar{\mu}_1(t) - \bar{\mu}_2(t))^2$. By analogy with the preceding example, we have $\mathbf{G}_2(t) = \mathbf{I}_2$ and $\mathbf{L}(t) = c_2(t)\,\mathbf{I}_2$. Further, from (4.11) and (4.7) one finds that $\mathbf{D} = (1 - a_2^2)\mathbf{I}_2$. It is shown in Appendix A that

$$\mathbf{S}_{2\infty} = \frac{1}{\bar{\mu}_1(t)} \mathbf{I}_2. \tag{4.42}$$

Hence, making the necessary substitutions in (4.23) and (4.24) we get

$$M_k(t) \sim \frac{(1 - a_2^2)\,(\bar{\mu}_1(t) - \bar{\mu}_2(t))}{(1 + (a_{12}^2 - a_{02}^2)t - \bar{\mu}_2(t))\,(1 - \bar{\mu}_2(t))\,\bar{\mu}_1^{1/2}(t)} \, \bar{\mu}_1^{-k/2}(t).$$

The derivation of $C(t)$ (in particular, that of $\mathbf{S}_{n_a\infty}(t)$ and $\Omega(t)$) can be quite tedious in more general cases. But, as far as the expressions for the ASN and OC function of SPRT are concerned, what is needed are certain properties of $\lambda(t)$ rather than a complete description of $C(t)$ and $\lambda(t)$. These properties will be investigated next.

4.2.2 Properties of $\lambda(t)$

The asymptotic approximation (4.22) reveals that the role of the moment generating function $m(t)$ in the independent case is taken over by $\lambda(t)$. The function $\lambda(t)$ is convex having a local minimum within Υ and attains value 1 at $t = 0$. The properties of $\lambda(t)$ that parallel those of $m(t)$ are stated in the following lemma.

Lemma 4.2 *The function $\lambda(t)$ has the following properties:*

i) $\lambda(0) = 1$,

ii) $\lambda'(0) = E\{z_k\}$,

iii) $\lambda''(0) > (\lambda'(0))^2$,

Chapter 4. Sequential Analysis of AR Processes 63

iv) If $\lambda'(0) \neq 0$, there exists a unique nonzero $t_1 \in \Upsilon$ such that $\lambda(t_1) = 1$.

Proof. From the distribution of $\mathbf{y}_a$ in (4.10), it is evident that $|\mathbf{V}| = |\mathbf{D}|$ for all $k > 0$. Therefore, (4.22) together with (4.14) implies

$$\begin{aligned}\lambda(0) &= \lim_{k\to\infty} \frac{M_k(0)}{M_{k-1}(0)} \\ &= \lim_{k\to\infty} \frac{|\mathbf{V}_k|^{-1/2}}{|\mathbf{V}_{k-1}|^{-1/2}} \\ &= 1\end{aligned}$$

where we have used a subscript to specify the size of $\mathbf{V}$. This proves i) and since $M_k(0) = 1$, it also implies $C(0) = 1$.

From (4.22), we can also get $E\{\mathcal{L}_k\} = kE\{z_k\} \sim k\lambda'(0) + C'(0)$. This means,

$$\lim_{k\to\infty} \frac{k\,E\{z_k\}}{k\,\lambda'(0)} = 1,$$

or simply $E\{z_k\} = \lambda'(0)$ which proves ii).

On the other hand, from (4.24), (4.16), (4.15) we note that

$$\lambda(t) = e^{v(t)} \tag{4.43}$$

where

$$v(t) = -\frac{1}{4\pi}\int_{-\pi}^{\pi} \ln\left[\sum_{n=-n_a}^{n_a} c_n(t)e^{jn\omega}\right] d\omega, \tag{4.44}$$

$$c_n(t) = \sum_{i=|n|}^{n_a} \left[(a_{1i}a_{1,i-|n|} - a_{0i}a_{0,i-|n|})t + a_i a_{i-|n|}\right] \tag{4.45}$$

with $a_0 = a_{10} = a_{00} = 1$. By taking derivatives of (4.43) we get

$$\begin{aligned}\lambda'(t) &= v'(t)\,\lambda(t) &(4.46)\\ \lambda''(t) &= \left(v''(t) + (v'(t))^2\right)\lambda(t). &(4.47)\end{aligned}$$

Since $c_n''(t) = 0$ we also have from (4.44)

$$v''(t) = \frac{1}{4\pi}\int_{-\pi}^{\pi} \left[\frac{\sum_{n=-n_a}^{n_a} c_n'(t)\,e^{jn\omega}}{\sum_{n=-n_a}^{n_a} c_n(t)\,e^{jn\omega}}\right]^2 d\omega > 0 \tag{4.48}$$

Hence from (4.46) and (4.47), $\lambda''(0) - (\lambda'(0))^2 = v''(0) > 0$ and iii) is proven.

To prove iv), we first note that (4.47) and (4.48) imply that $\lambda''(t) > 0$ for all $t \in \Upsilon$. Consider the case where Υ has finite endpoints. (Otherwise, the proof follows immediately from the convexity of $\lambda(t)$.) As t approaches the endpoints of Υ (which contains 0) the determinant of $(\mathbf{Q}_1^T\mathbf{Q}_1 - \mathbf{Q}_0^T\mathbf{Q}_0)t + \mathbf{V}$ will tend to zero. Hence, by Lemma 4.1 there will be both positive and negative values of t for which $\mathcal{G}(t) < 1$, or equally $\lambda(t) > 1$. Since $\lambda(t)$ is a convex function and $\lambda(0) = 1$, if $\lambda'(0) \neq 0$ there must exist a unique $t_1 \in \Upsilon$ such that $\lambda(t_1) = 1$. □

Remark From the convexity of $\lambda(t)$, we can conclude that $t_1 = 0$ if $\lambda'(0) = 0$. In fact, because of the continuity of $\lambda'(t)$ in Υ it follows that

$$t_1 \to 0 \qquad \text{as} \quad \lambda'(0) \to 0. \qquad (4.49)$$

By virtue of ii) and iii) in Lemma 4.2, it follows from (4.22) that $\mathcal{L}_k$ asymptotically has a normal distribution with mean $k\lambda'(0)$ and variance $k(\lambda''(0) - (\lambda'(0))^2)$. Therefore, by denoting the termination time of SPRT by n we can write

$$\Pr\{n > k\} \leq \Pr\{\alpha < \mathcal{L}_k < \beta\}$$
$$= \operatorname{erf}\left(\frac{\beta - k\lambda'(0)}{[k(\lambda''(0) - (\lambda'(0))^2)]^{1/2}}\right) - \operatorname{erf}\left(\frac{\alpha - k\lambda'(0)}{[k(\lambda''(0) - (\lambda'(0))^2)]^{1/2}}\right).$$

Hence, $\Pr\{n > N\} \to 0$ as $k \to \infty$. This, in fact, proves the termination property of SPRT in the AR case:

Theorem 4.2 *The SPRT among the hypotheses $\mathcal{H}_0$ and $\mathcal{H}_1$ given by (4.3) and (4.4), respectively, will terminate with probability 1.*

The next lemma specifies the values of t_1 in iv) of Lemma 4.2, when either $\mathcal{H}_0$ or $\mathcal{H}_1$ is true.

Lemma 4.3 *The root of equation $\lambda(t) = 1$ is*

$$\begin{array}{ll} t_1 = 1 & \text{if } \mathcal{H}_0 \text{ is true,} \\ t_1 = -1 & \text{if } \mathcal{H}_1 \text{ is true.} \end{array} \qquad (4.50)$$

Proof. First by using (4.45) in (4.15) we note that

$$g(e^{j\omega}, t) = (|A_1(e^{j\omega})|^2 - |A_0(e^{j\omega})|^2)t + |A(e^{j\omega})|^2,$$

where A_0 and A_1 are the A polynomials corresponding to the hypotheses $\mathcal{H}_0$ and $\mathcal{H}_1$, respectively. Hence, under $\mathcal{H}_0$, $g(e^{j\omega}, 1) = |A_1(e^{j\omega})|^2$ and, under $\mathcal{H}_1$, $g(e^{j\omega}, -1) = |A_0(e^{j\omega})|^2$. Therefore, in view of (4.24) and (4.16), to prove the lemma it is sufficient if we show

$$\int_{-\pi}^{\pi} \ln |A(e^{j\omega})|^2 \, d\omega = 0 \tag{4.51}$$

for any polynomial A as in (4.2). To do this, $A(z)$ can be factorized using its monicity

$$A(z) = (1 - \zeta_1^{-1} z)(1 - \zeta_2^{-1} z) \cdots (1 - \zeta_{n_a}^{-1} z) \tag{4.52}$$

where ζ_i ($i = 1, \ldots, n_a$) are the roots of $A(z)$ which were assumed to be outside the unit circle. We can rewrite the integral in (4.51)

$$\int_{-\pi}^{\pi} \ln |A(e^{j\omega})|^2 \, d\omega = \sum_{i=1}^{n_a} I_i \tag{4.53}$$

where $I_i = \int_{-\pi}^{\pi} \ln |1 - \zeta_i^{-1} e^{j\omega}|^2 \, d\omega$ or, rearranging,

$$\begin{aligned} I_i &= \int_{-\pi}^{\pi} \ln \left[1 + |\zeta_i|^{-2} - 2 \Re\{\zeta_i^{-1} e^{j\omega}\} \right] d\omega \\ &= \int_{-\pi}^{\pi} \ln \left[1 + |\zeta_i|^{-2} - 2|\zeta_i|^{-1} \cos(\omega - \omega_{\zeta_i}) \right] d\omega \end{aligned}$$

where $\Re$ denotes the real part and ω_{ζ_i} denotes the angle of ζ_i. Because of periodicity and evenness of cosine function,

$$I_i = 2 \int_0^{\pi} \ln \left[1 + |\zeta_i|^{-2} - 2|\zeta_i|^{-1} \cos \omega \right] d\omega.$$

Noting that $1 + |\zeta_i|^{-2} > 2|\zeta_i|^{-1}$ for any ζ_i which is not on the unit circle, and using the integral formula (4.31), we obtain

$$I_i = 2\pi \ln \frac{1 + |\zeta_i|^{-2} + \sqrt{(1 - |\zeta_i|^{-2})^2}}{2}$$

and further, since $|\zeta_i|^{-1} < 1$, $I_i = 0$. By (4.53), this proves (4.51) and, hence, the lemma. □

4.3 An Analogue of Fundamental Identity

In this section, we prove an analogue of the Fundamental Identity of sequential analysis corresponding to the AR case where the samples are obtained from (4.1).

Theorem 4.3 *Let n be the smallest integer such that the cumulative sum $\mathcal{L}_n$ defined by (3.2) and (4.6) is outside (α, β), $(\alpha < 0 < \beta)$. Then, for values of t such that $\lambda(t)$ exists and $\lambda(t) \geq 1$,*

$$E\left\{e^{\mathcal{L}_n t}\lambda^{-n}(t)\bar{C}(t,\hat{\mathbf{y}}_n)\right\} = 1 \qquad (4.54)$$

where

$$\begin{aligned}
\bar{C}(t,\hat{\mathbf{y}}_n) &= \frac{1}{|\mathbf{D}|^{1/2}}\exp\left\{-\frac{1}{2\sigma^2}\hat{\mathbf{y}}_n^T[\mathbf{G}_1(t)-\mathbf{D}-\mathbf{L}(t)\mathbf{S}_{n_a\infty}(t)\mathbf{L}^T(t)]\hat{\mathbf{y}}_n\right\} \\
&\quad \times |\mathbf{G}_1(t) - \mathbf{L}(t)\mathbf{S}_{n_a\infty}(t)\mathbf{L}^T(t)|^{1/2}
\end{aligned} \qquad (4.55)$$

and $\hat{\mathbf{y}}_n = [y(n-n_a+1),\ldots,y(n)]^T$.

Proof. Let $K > N > 0$ and $P_N = \Pr\{n \leq N\}$. Then we have

$$\begin{aligned}
M_K(t) &= E\left\{e^{\mathcal{L}_K t}\right\} \\
&= P_N E\{\exp[\mathcal{L}_n t + (\mathcal{L}_K - \mathcal{L}_n)t] \mid n \leq N\} \\
&\quad + (1 - P_N) E\{\exp[\mathcal{L}_N t + (\mathcal{L}_K - \mathcal{L}_N)t] \mid n > N\}.
\end{aligned} \qquad (4.56)$$

Since $\mathcal{L}_K - \mathcal{L}_n$ $(K > n)$ depends on $\mathcal{L}_n$ only through $y(n),\ldots,y(n-n_a+1)$, it follows that

$$E\{\exp[\mathcal{L}_n t + (\mathcal{L}_K - \mathcal{L}_n)t] \mid n \leq N\} = E\{e^{\mathcal{L}_n t} M_{K-n}(t \mid \hat{\mathbf{y}}_n) \mid n \leq N\} \qquad (4.57)$$

where $M_{K-n}(t \mid \hat{\mathbf{y}}_n) = E\{\exp[(z_{n+1} + \cdots + z_K)t] \mid \hat{\mathbf{y}}_n\}$ with $K > n$. It is proven in Appendix B that for large values of $K - n$

$$M_{K-n}(t \mid \hat{\mathbf{y}}_n) \sim D(t,\hat{\mathbf{y}}_n)\lambda^{K-n}(t) \qquad (4.58)$$

with

$$D(t,\hat{\mathbf{y}}_n) = \frac{\exp\left\{-\frac{1}{2\sigma^2}\hat{\mathbf{y}}_n^T[\mathbf{G}_1(t)-\mathbf{D}-\mathbf{L}(t)\mathbf{S}_{n_a\infty}(t)\mathbf{L}^T(t)]\hat{\mathbf{y}}_n\right\}\mathcal{G}^{n_a/2}(t)}{|\mathbf{G}_2(t)-\mathbf{L}(t)\mathbf{S}_{n_a\infty}(t)\mathbf{L}^T(t)|^{1/2}\Omega^{1/2}(t)}. \qquad (4.59)$$

Similar to (4.57), we have

$$E\{\exp[\mathcal{L}_N t + (\mathcal{L}_K - \mathcal{L}_N)t] \mid n > N\} = E\{e^{\mathcal{L}_N t} M_{K-N}(t \mid \hat{\mathbf{y}}_N) \mid n > N\}. \quad (4.60)$$

Using (4.57) and (4.60) in (4.56) and dividing both sides of (4.56) by $M_K(t)$, one gets

$$P_N E\left\{e^{\mathcal{L}_n t} \frac{M_{K-n}(t \mid \hat{\mathbf{y}}_n)}{M_K(t)} \mid n \leq N\right\} + (1 - P_N) E\left\{e^{\mathcal{L}_N t} \frac{M_{K-N}(t \mid \hat{\mathbf{y}}_N)}{M_K(t)} \mid n > N\right\} = 1. \quad (4.61)$$

First we let $K \to \infty$ keeping N fixed. From (4.22), for values of t such that $\lambda(t) \geq 1$, $M_K^{-1}(t)$ will remain finite; therefore, so will the second term on the left hand side of (4.61). On the other hand, if $n \leq N$

$$\begin{aligned}\lim_{K \to \infty} \frac{M_{K-n}(t \mid \hat{\mathbf{y}}_n)}{M_K(t)} &= \frac{D(t, \hat{\mathbf{y}}_n) \lambda^{K-n}(t)}{C(t) \lambda^K(t)} \\ &= \bar{C}(t, \hat{\mathbf{y}}_n) \lambda^{-n}(t).\end{aligned}$$

$\bar{C}(t, \hat{\mathbf{y}}_n)$ being as in (4.55). As $N \to \infty$, $P_N \to 1$ from Theorem 4.2 and hence we obtain from (4.61)

$$E\left\{e^{\mathcal{L}_n t} \lambda^{-n}(t) \bar{C}(t, \hat{\mathbf{y}}_n)\right\} = 1.$$

□

Remark Note that although the relation (4.22) which plays the key role in the proof of Theorem 4.3 is only an asymptotic approximation the Fundamental Identity in (4.54) holds with exact equality. In contrast to the i.i.d. case, it involves the values of some samples taken just prior to the termination of the test.

4.4 ASN and OC Function

Having established the Fundamental Identity for the AR case, we can now proceed as in the previous chapter to obtain the ASN and OC function of SPRT.

The ASN can be found from (4.54) by taking the derivatives of both sides and putting $t = 0$. Assuming that the order of differentiation and expectation operators can be exchanged, one gets, since $\lambda(0) = 1$,

$$E\left\{\mathcal{L}_n \bar{C}(0,\hat{\mathbf{y}}_n) - n\,\lambda'(0) + \bar{C}'(0,\hat{\mathbf{y}}_n)\right\} = 1. \tag{4.62}$$

In fact, $\lambda(0) = 1$ also implies, by (4.22) and (4.58), that $C(0) = D(0,\hat{\mathbf{y}}_n) = 1$. Hence, $\bar{C}(0,\hat{\mathbf{y}}_n) = 1$. So, solving for $E\{n\}$ in (4.62), we obtain

$$E\{n\} = \frac{E\{\mathcal{L}_n\} + E\left\{\bar{C}'(0,\hat{\mathbf{y}}_n)\right\}}{E\{z_k\}} \tag{4.63}$$

for $E\{z_k\} = \lambda'(0) \neq 0$. For the case when $\lambda'(0) = 0$, we differentiate (4.54) twice and set $t = 0$ to get

$$E\left\{\mathcal{L}_n^2 + 2\mathcal{L}_n\,\bar{C}'(0,\hat{\mathbf{y}}_n) + \bar{C}''(0,\hat{\mathbf{y}}_n) - n\,\lambda''(0)\right\} = 0$$

and solve for $E\{n\}$ as

$$E\{n\} = \frac{E\{\mathcal{L}_n^2\} + 2E\left\{\mathcal{L}_n \bar{C}'(0,\bar{\mathbf{y}}_n)\right\} + E\left\{\bar{C}''(0,\hat{\mathbf{y}}_n)\right\}}{\lambda''(0)}. \tag{4.64}$$

On the other hand, to find an expression for the OC function we evaluate the Fundamental Identity at $t = t_1$. We have from from (4.54)

$$P_\alpha(\boldsymbol{\theta})e^{\alpha t_1}E\{\bar{C}(t_1,\hat{\mathbf{y}}_n)\mid \mathcal{L}_n = \alpha\} + (1 - P_\alpha(\boldsymbol{\theta}))e^{\beta t_1}E\{\bar{C}(t_1,\hat{\mathbf{y}}_n)\mid \mathcal{L}_n = \beta\} \approx 1. \tag{4.65}$$

Note that in (4.65) we ignored the excess of the $\mathcal{L}_n$ beyond the thresholds α and β. The OC function follows from (4.65) as

$$P_\alpha(\boldsymbol{\theta}) \approx \frac{1 - e^{\beta t_1}E\{\bar{C}(t_1,\hat{\mathbf{y}}_n)\mid \mathcal{L}_n = \beta\}}{e^{\alpha t_1}E\{\bar{C}(t_1,\hat{\mathbf{y}}_n)\mid \mathcal{L}_n = \alpha\} - e^{\beta t_1}E\{\bar{C}(t_1,\hat{\mathbf{y}}_n)\mid \mathcal{L}_n = \beta\}}. \tag{4.66}$$

Unfortunately, the formulae (4.63), (4.64) and (4.66) involve the values of observations made just prior to the termination of the test, and these are generally not available a priori. Nevertheless, one can obtain useful approximations if $\hat{\mathbf{y}}_n$ is assumed to have the stationary distribution of n_a consecutive samples. This implies

$$f(\hat{\mathbf{y}}_n) = \frac{|\mathbf{D}|^{1/2}}{(\sigma\sqrt{2\pi})^{n_a}} \exp\left\{-\frac{1}{2\sigma^2}\hat{\mathbf{y}}_n^T \mathbf{D} \hat{\mathbf{y}}_n\right\}. \tag{4.67}$$

Chapter 4. Sequential Analysis of AR Processes

From (4.67) and (4.55) it follows that

$$\begin{aligned}
E\{\bar{C}(t,\hat{\mathbf{y}}_n)\} &= |\mathbf{G}_1(t) - \mathbf{L}(t)\mathbf{S}_{n_a\infty}(t)\mathbf{L}^T(t)|^{1/2}(\sigma\sqrt{2\pi})^{-n_a} \\
&\quad \times \int_{-\infty}^{\infty} \exp\left\{-\frac{1}{2\sigma^2}\hat{\mathbf{y}}_n^T[\mathbf{G}_1(t) - \mathbf{D} - \mathbf{L}(t)\mathbf{S}_{n_a\infty}(t)\mathbf{L}^T(t)]\hat{\mathbf{y}}_n\right\} \\
&\quad \times \exp\left\{-\frac{1}{2\sigma^2}\hat{\mathbf{y}}_n^T\mathbf{D}\hat{\mathbf{y}}_n\right\} d\hat{\mathbf{y}}_n \\
&= 1.
\end{aligned}$$

Therefore the expected value of all derivatives of $\bar{C}(t,\hat{\mathbf{y}}_n)$ vanishes everywhere. By neglecting the overshoot of $\mathcal{L}_n$ beyond the thresholds we obtain the same approximate expression for ASN as in the i.i.d. case, namely,

$$E\{n\} \approx \frac{\alpha P_\alpha(\boldsymbol{\theta}) + \beta(1 - P_\alpha(\boldsymbol{\theta}))}{E\{z_k\}} \tag{4.68}$$

for $\lambda'(0) \neq 0$. On the other hand, the formula for the case $\lambda'(0) = 0$ reduces approximately to

$$E\{n\} \approx \frac{\alpha^2 P_\alpha(\boldsymbol{\theta}) + \beta^2(1 - P_\alpha(\boldsymbol{\theta}))}{\lambda''(0)}. \tag{4.69}$$

Setting $E\{\bar{C}(t_1,\hat{\mathbf{y}}_n) \mid \mathcal{L}_n = \alpha\} = E\{\bar{C}(t_1,\hat{\mathbf{y}}_n) \mid \mathcal{L}_n = \beta\} \approx 1$, the operating characteristics of the test will be approximately

$$P_\alpha(\boldsymbol{\theta}) \approx \frac{1 - e^{\beta t_1}}{e^{\alpha t_1} - e^{\beta t_1}} \tag{4.70}$$

which also has the same form as in the i.i.d. case. Note that, by (4.49), for the value of $\boldsymbol{\theta}$ for which $t_1 = 0$, (4.70) must be evaluated as a limit. That is,

$$\begin{aligned}
P_\alpha(\boldsymbol{\theta}) &\approx \lim_{t_1 \to 0} \frac{1 - e^{\beta t_1}}{e^{\alpha t_1} - e^{\beta t_1}} \\
&= \frac{\beta}{\beta - \alpha} \qquad \text{for} \quad \lambda'(0) = 0.
\end{aligned} \tag{4.71}$$

4.5 The Exact Log Likelihood Ratio

In Section 4, the analysis was based on the conditional log likelihood ratio. The same line of thought can be followed also to derive the Fundamental Identity and sampling properties of SPRT when the exact log likelihood ratio, which is calculated

Chapter 4. Sequential Analysis of AR Processes 70

recursively using the scores in (4.5), is employed. The analysis of this case is given in detail by Kerestecioğlu and Zarrop (1990). To avoid repetitions here, only some intermediate steps which differ from the conditional case, will be indicated.

The joint probability distribution of $\mathbf{y}_k$ is as in (4.10) save that $\mathbf{y}_a$ is replaced by $\mathbf{y}_k$. Therefore, if one uses the exact log likelihood ratio, $\mathcal{L}_k = f_1(\mathbf{y}_k)/f_0(\mathbf{y}_k)$, referring to (4.13) and (4.14), its moment generating function will be

$$M_k(t) = \frac{|\mathbf{D}_1|^{t/2}|\mathbf{D}|^{1/2}}{|\mathbf{D}_0|^{t/2}(\sigma\sqrt{2\pi})^k} \int_{-\infty}^{\infty} \exp\left\{-\frac{1}{2\sigma^2}\mathbf{y}_k^T[(\mathbf{V}_1 - \mathbf{V}_0^T)t + \mathbf{V}]\mathbf{y}_k\right\} d\mathbf{y}_k.$$

$$= \frac{|\mathbf{D}_1|^{t/2}|\mathbf{D}|^{1/2}}{|\mathbf{D}_0|^{t/2}|(\mathbf{V}_1 - \mathbf{V}_0)t + \mathbf{V}|^{1/2}}. \tag{4.72}$$

where $\mathbf{D}_i$ and $\mathbf{V}_i$ correspond to $\mathbf{D}$ and $\mathbf{V}$ matrices under hypotheses $\mathcal{H}_i$ ($i = 0, 1$), respectively.

The $k \times k$ matrix $(\mathbf{V}_1 - \mathbf{V}_0)t + \mathbf{V}$ is doubly symmetric. Therefore, in the notation of Section 2 the role of $\mathbf{G}_1(t)$ is replaced by $\mathbf{G}_2(t)$. Using Lemma 4.1 and (4.72), the function $C(t)$ in the asymptotic relation (4.22) turns out to be

$$C(t) = \frac{|\mathbf{D}_1|^{t/2}|\mathbf{D}|^{1/2}\mathcal{G}^{n_a}(t)}{|\mathbf{D}_0|^{t/2}|\mathbf{G}_2(t) - \mathbf{L}(t)\mathbf{S}_{n_a\infty}(t)\mathbf{L}^T(t)|\,\Omega^{1/2}(t)} \tag{4.73}$$

whereas we still have $\lambda(t) = \mathcal{G}^{-1/2}(t)$.

Example 4.1 (continued) Consider the first order model in (4.25). The function $C(t)$ for this case can be found by noting $\mathbf{D}_0 = 1 - a_{01}^2$ and $\mathbf{D}_1 = 1 - a_{11}^2$ and using (4.26), (4.32), (4.36) (4.37) and (4.39) in (4.73) as

$$C(t) = \left(\frac{1 - a_{11}^2}{1 - a_{01}^2}\right)^{t/2} \frac{(1 - a_1^2)^{1/2}\,(\mu_1(t) - \mu_2(t))^{1/2}\,\mu_1^{1/2}(t)}{1 - \mu_2(t)}$$

This expression is the same as found by Phatarfod (1971) for this Gauss-Markov case by using a recursive formulation for expansion of the determinant of the tridiagonal $(\mathbf{V}_1 - \mathbf{V}_0)t + \mathbf{V}$ matrix.

Proceeding along the same lines as in Section 3 one arrives at the Fundamental Identity (4.54). But this time $\bar{C}(t, \hat{\mathbf{y}}_n) = D(t, \hat{\mathbf{y}}_n)/C(t)$ follows from (4.73) and

Chapter 4. Sequential Analysis of AR Processes

(4.59) as

$$\bar{C}(t,\hat{\mathbf{y}}_n) = \left[\frac{|\mathbf{D}_0|^t \, |\mathbf{G}_2(t) - \mathbf{L}(t)\mathbf{S}_{n_a\infty}(t)\mathbf{L}^T(t)|}{|\mathbf{D}_1|^t \, |\mathbf{D}| \, \mathcal{G}^{n_a}(t)}\right]^{1/2}$$
$$\times \exp\left\{-\frac{1}{2\sigma^2}\hat{\mathbf{y}}_n^T[\mathbf{G}_1(t) - \mathbf{D} - \mathbf{L}(t)\mathbf{S}_{n_a\infty}(t)\mathbf{L}^T(t)]\hat{\mathbf{y}}_n\right\}. \quad (4.74)$$

To find an approximation for the exact formula for the ASN in (4.63) one has to approximate $E\{\bar{C}(t,\hat{\mathbf{y}}_n)\}$, with $\bar{C}(t,\hat{\mathbf{y}}_n)$ being as in (4.74). As in the conditional log likelihood ratio case we assume that $\hat{\mathbf{y}}_n$ has the stationary distribution (4.67). Taking the expectation of $\bar{C}(t,\hat{\mathbf{y}}_n)$ and noting that $\mathcal{G}^{-n_a/2}(t) = \lambda^{n_a}(t)$, it follows that

$$E\{\bar{C}(t,\hat{\mathbf{y}}_n)\} = \left[\frac{|\mathbf{D}_0|^t \, |\mathbf{G}_2(t) - \mathbf{L}(t)\mathbf{S}_{n_a\infty}(t)\mathbf{L}^T(t)|}{|\mathbf{D}_1|^t \, |\mathbf{G}_1(t) - \mathbf{L}(t)\mathbf{S}_{n_a\infty}(t)\mathbf{L}^T(t)|}\right]^{1/2} \lambda^{n_a}(t).$$

Recall that $\mathbf{G}_1(t)$ and $\mathbf{G}_2(t)$ were defined as the n_a-th leading submatrices of $(\mathbf{Q}_1^T\mathbf{Q}_1 - \mathbf{Q}_0^T\mathbf{Q}_0)t + \mathbf{V}$. Hence, in view of (4.11), $\mathbf{G}_1(t) = \mathbf{G}_2(t) + (\mathbf{D}_1 - \mathbf{D}_0)t$. Therefore, under the close hypotheses assumption

$$E\{\bar{C}(t,\hat{\mathbf{y}}_n)\} \approx \lambda^{n_a}(t) \quad (4.75)$$

and, further, $\bar{C}'(0,\hat{\mathbf{y}}_n) \approx n_a \lambda'(0) = n_a E\{z\}$. The approximate expression for the ASN turns out to be from (4.63)

$$E\{n\} \approx \frac{\alpha P_\alpha(\boldsymbol{\theta}) + \beta(1 - P_\alpha(\boldsymbol{\theta}))}{E\{z_k\}} + n_a.$$

Evidently, this is the same average sample number as one would get by only observing the initial values $\{y(1),\ldots,y(n_a)\}$ without adding the increments $\{z_1,\ldots,z_{n_a}\}$ to the log likelihood ratio and thereafter using the log likelihood ratio conditioned on these samples to conduct the test.

On the other hand, the assumption that $\hat{\mathbf{y}}_n$ is distributed according to (4.67) when $\mathcal{L}_n = \alpha$ or $\mathcal{L}_n = \beta$ leads to $E\{\bar{C}(t_1,\hat{\mathbf{y}}_n) \mid \mathcal{L}_n = \alpha\} = E\{\bar{C}(t_1,\hat{\mathbf{y}}_n) \mid \mathcal{L}_n = \beta\} \approx \lambda^{n_a}(t_1) = 1$ by (4.75), and hence validates the approximation

$$P_\alpha(\boldsymbol{\theta}) \approx \frac{1 - e^{\beta t_1}}{e^{\alpha t_1} - e^{\beta t_1}}$$

for the OC function in (4.66) in the exact log likelihood ratio case as well.

4.6 Simulation Examples

In this section, we test the validity of the approximate formulae for the ASN and the OC function in the AR case.

Example 4.3 First, let us consider the first order autoregressive process $A(q^{-1})y(k) = \epsilon(k)$ where $A(q^{-1}) = 1 + a_1 q^{-1}$ with two hypotheses about the value of the parameter a_1, which are

$$\begin{aligned}\mathcal{H}_0: \ & a_1 = a_{01} = 0.4 \\ \mathcal{H}_1: \ & a_1 = a_{11} = 0.5.\end{aligned} \quad (4.76)$$

The variance of the zero mean white Gaussian noise $\epsilon(k)$ is assumed to be 1. With the application of (4.6), $\mathcal{L}_k$ is to be computed by adding the scores

$$\begin{aligned}z_k &= \frac{1}{2}(y(k) + a_{01}y(k-1))^2 - \frac{1}{2}(y(k) + a_{11}y(k-1))^2 \\ &= \frac{1}{2}(a_{01}^2 - a_{11}^2)y^2(k-1) - (a_{01} - a_{11})y(k)y(k-1) \end{aligned} \quad (4.77)$$

at each sampling instant. Adopting $\epsilon_1 = \epsilon_2 = 0.1$ as the error probabilities under $\mathcal{H}_0$ and $\mathcal{H}_1$, respectively, the thresholds of the test turn out to be $\alpha = -2.197$ and $\beta = 2.197$.

To find an approximate expression for the OC we need $t_1(a_1)$ where $\lambda(t_1) = 1$. The function $\lambda(t)$ corresponding to this first order case has already been found in Example 2 as $\mu_1^{-1/2}(t)$, where $\mu_1(t)$ is given by (4.29). By using (4.28) in (4.29) and solving $\mu_1(t_1) = 1$ we obtain

$$t_1 = \frac{a_{11} + a_{01} - 2a_1}{a_{11} - a_{01}} = 9 - 20a_1. \quad (4.78)$$

It is interesting to note that the expression of t_1 in (4.78) has the same form as that of the SPRT for the mean of a Gaussian distribution in the i.i.d. case. By using (4.78) in (4.70) and noting that $\alpha = -\beta$, we get

$$P_\alpha(a_1) \approx \frac{e^{\beta t_1(a_1)} - 1}{2\sinh(\beta t_1(a_1))}. \quad (4.79)$$

Chapter 4. Sequential Analysis of AR Processes   73

On the other hand, to find the ASN we first note that the correlations of $y(k)$ can be obtained by

$$E\{y(k)\,y(k-l)\} = \frac{1}{2\pi}\int_{-\pi}^{\pi}\frac{e^{jl\omega}}{|A(e^{j\omega})|^2}\,d\omega$$

which yields for $l = 0, 1$,

$$E\{y^2(k)\} = \frac{1}{1-a_1^2} \qquad E\{y(k)\,y(k-1)\} = \frac{a_1}{1-a_1^2}. \tag{4.80}$$

The average value of the increments of the log likelihood ratio can be found by taking expectations in (4.77) and using (4.80) as

$$\begin{aligned} E\{z_k \mid a_1\} &= \frac{1}{2(1-a_1^2)}[a_{01}^2 - a_{11}^2 + 2a_1(a_{11} - a_{01})] \\ &= \frac{0.1(a_1 - 0.45)}{1-a_1^2}. \end{aligned} \tag{4.81}$$

Therefore, for $a_1 \neq 0.45$ the average sample number is approximately

$$E\{n \mid a_1\} \approx \frac{(1-a_1^2)\,(1-\cosh(\beta t_1 a_1))}{\sinh(\beta t_1 a_1)}\beta. \tag{4.82}$$

Figure 4.1 displays the ASN (4.82) as a function of the true parameter value in $[0, 1)$. It also shows the results obtained in simulations based on 5000 runs for various values of a_1. Note that, as was found in the i.i.d. case, the estimated values of the ASN are slightly larger than those given by the approximate formula which is basically due to neglecting the overshoot of $\mathcal{L}_n$ over the thresholds.

The OC function (4.79) is shown in Figure 4.2 together with the results obtained from the simulations. The approximation in (4.79) proved to be quite close to estimated values. Note that the estimated probabilities of making errors under $\mathcal{H}_0$ and $\mathcal{H}_1$ are in fact slightly less than the specified values (namely, $\epsilon_1 = \epsilon_2 = 0.1$).

It is clear from the Figures 4.1 and 4.2 that SPRT about the hypotheses in (4.76) can be applied successfully to test the composite hypotheses $\mathcal{H}_0$: $a_1 < a_{01}$ and $\mathcal{H}_1$: $a_1 > a_{11}$.

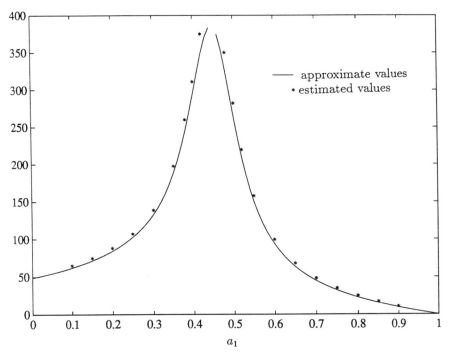

Figure 4.1: ASN of SPRT in testing first order AR parameter

Example 4.4 As another example we shall consider a second order process, i.e. one with $A(q^{-1}) = 1 + a_1 q^{-1} + a_2 q^{-2}$ and the hypotheses about $\boldsymbol{\theta} = [a_1 \ a_2]^T$,

$$\mathcal{H}_0 : \ \boldsymbol{\theta} = \boldsymbol{\theta}_0 = [0.6 \ \ 0.25]^T$$
$$\mathcal{H}_1 : \ \boldsymbol{\theta} = \boldsymbol{\theta}_1 = [0.7 \ \ 0.325]^T.$$

Under $\mathcal{H}_0(\mathcal{H}_1)$ the polynomial $A(z^{-1})$ has zeros $-0.3 \pm 0.4j(-0.35 \pm 0.45j)$. The variance of $\epsilon(k)$ is again assumed to be 1. The increments z_k will be in this case

$$\begin{aligned}
z_k &= \frac{1}{2}(y(k) + a_{01} y(k-1) + a_{02} y(k-2))^2 - \\
& \quad \frac{1}{2}(y(k) + a_{11} y(k-1) + a_{12} y(k-2))^2 \\
&= (a_{01}^2 - a_{11}^2) y^2(k-1) + (a_{02}^2 - a_{12}^2) y^2(k-2) + 2(a_{01} - a_{11}) y(k) y(k-1) \\
& \quad + 2(a_{02} - a_{12}) y(k) y(k-2) + 2(a_{01} a_{02} - a_{11} a_{12}) y(k-1) y(k-2) \quad (4.83)
\end{aligned}$$

The simulations are done to evaluate the ASN and error probabilities for three different threshold combinations. Table 4.1 compares the values estimated via the

Chapter 4. Sequential Analysis of AR Processes

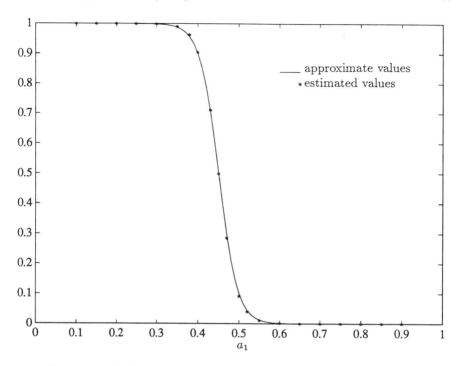

Figure 4.2: OC function of SPRT in testing first order AR parameter

simulations and the approximate nominal values obtained by using the formulae (4.68) and (4.70). It is seen that that the approximations for ASN and OC are valid save for the slight increase in ASN similar to that in the i.i.d. or first order AR cases.

4.7 Conclusions

We have investigated the performance of SPRT in deciding between the hypotheses about the parameters of a stationary autoregressive process driven by white Gaussian noise. Towards this goal, first, the Fundamental Identity of sequential analysis is proven for the AR case. Then, two important quantities describing the performance of the test, namely the ASN and OC function followed from the Fundamental Identity in a way similar to the i.i.d. case. In this analysis we observed that the role of the moment generating function of scores z_k in the independent case is taken over

$\alpha = -2.197, \beta = 2.197$	ϵ_1	ϵ_2	$E\{n \mid \mathcal{H}_0\}$	$E\{n \mid \mathcal{H}_1\}$
Nominal	0.100	0.100	301	294
Estimated	0.090	0.097	324	308

$\alpha = -2.251, \beta = 2.890$	ϵ_1	ϵ_2	$E\{n \mid \mathcal{H}_0\}$	$E\{n \mid \mathcal{H}_1\}$
Nominal	0.050	0.100	342	398
Estimated	0.048	0.091	359	420

$\alpha = -2.890, \beta = 2.251$	ϵ_1	ϵ_2	$E\{n \mid \mathcal{H}_0\}$	$E\{n \mid \mathcal{H}_1\}$
Nominal	0.100	0.050	407	340
Estimated	0.087	0.046	431	350

Table 4.1: Performance of SPRT in testing second order AR parameters by the function $\lambda(t)$.

The results in this chapter reduce for $n_a = 1$ to those obtained by Phatarfod (1971) for the Gauss-Markov processes. It is also interesting to note that the approximate expressions for ASN and OC are very similar to those for the case where the observations are i.i.d.. In contrast to the i.i.d. case the approximate formulae (4.68) and (4.70) involve an additional assumption about the distribution of the samples just prior to the termination of the test (namely, $\hat{\mathbf{y}}_n$). But the simulations indicate that this does not degrade the quality of approximations.

In this chapter we restricted ourselves to the AR case. However, since any autoregressive moving average (ARMA) process can be approximated closely enough by an AR process with finite order, it is believed that the results would hold for ARMA processes as well. Therefore, in the next chapter, where the application of SPRT to the problem of detecting changes which may occur at an unknown sampling time will be discussed, the author will feel free to use models having moving average as well as autoregressive parts.

Appendix 4.A Derivation of Equation (4.42)

In this appendix we show that, as $k \to \infty$, the second leading submatrix of $\mathbf{T}_k^{-1}$ in Example 4.2 is given by (4.42). For notational simplicity, the argument t of some quantities will be omitted. The matrix $\mathbf{T}_k$ is the Toeplitz matrix associated with the nearly Toeplitz matrix in (4.40) and it can be written as $\mathbf{T}_k = c_0 \bar{\mathbf{T}}_k$ where $\bar{\mathbf{T}}_k$ is Toeplitz with elements

$$\bar{c}_{ij} = \begin{cases} 1 & \text{if } i = j \\ \dfrac{c_2}{c_0} & \text{if } |i-j| = 2 \\ 0 & \text{otherwise} \end{cases}$$

for $i, j = 1, \ldots, k$. Let us denote the elements of $\bar{\mathbf{T}}_k^{-1}$ by w_{ij}. Then $\{w_{ik}\}_{i=1}^k$ satisfy the system of linear equations

$$0 = w_{1k} + \frac{c_2}{c_0} w_{3k} \qquad (4.84)$$

$$0 = w_{2k} + \frac{c_2}{c_0} w_{4k} \qquad (4.85)$$

$$0 = \frac{c_2}{c_0} w_{k-3,k} + w_{k-1,k} \qquad (4.86)$$

$$1 = \frac{c_2}{c_0} w_{k-2,k} + w_{kk} \qquad (4.87)$$

$$0 = \frac{c_2}{c_0} w_{i-2,k} + w_{ik} + \frac{c_2}{c_0} w_{i+2,k} \qquad i = 3, \ldots, k-2. \qquad (4.88)$$

The characteristic equation of the difference equation in (4.88) is

$$z^4 + \frac{c_0}{c_2} z^2 + 1 = 0$$

which has roots $\zeta = \sqrt{-\bar{\mu}_1/c_2}$, ζ^{-1}, $-\zeta$, and $-\zeta^{-1}$. Therefore, w_{ik} can be written in the form

$$w_{ik} = C_1 \zeta^i + C_2 \zeta^{-i} + C_3 (-\zeta)^i + C_4 (-\zeta)^{-i}. \qquad (4.89)$$

The constants $C_1, \ldots, C_4$ are to be determined from (4.84)–(4.87). Assume that k is even, that is, $(-\zeta)^k = \zeta^k$. By substituting (4.84) – (4.87) into (4.88) and using

Chapter 4. Sequential Analysis of AR Processes

the relations $1 + \zeta^2 c_2/c_0 = \bar{\mu}_2/c_0$ and $1 + \zeta^{-2} c_2/c_0 = \bar{\mu}_1/c_0$ we will end up with

$$\frac{1}{c_0}\begin{bmatrix} \mathbf{A}_{11} & \mathbf{A}_{12} \\ \mathbf{A}_{21} & \mathbf{A}_{22} \end{bmatrix} \begin{bmatrix} C_1 \\ C_3 \\ C_2 \\ C_4 \end{bmatrix} = \begin{bmatrix} 0 \\ 0 \\ 0 \\ 1 \end{bmatrix}$$

where

$$\mathbf{A}_{11} = \bar{\mu}_2 \begin{bmatrix} \zeta & -\zeta \\ \zeta^2 & \zeta^2 \end{bmatrix}, \quad \mathbf{A}_{12} = \bar{\mu}_1 \begin{bmatrix} \zeta^{-1} & -\zeta^{-1} \\ \zeta^{-2} & \zeta^{-2} \end{bmatrix},$$

$$\mathbf{A}_{21} = \bar{\mu}_1 \begin{bmatrix} \zeta^{k-1} & -\zeta^{k-1} \\ \zeta^k & \zeta^k \end{bmatrix}, \quad \mathbf{A}_{22} = \bar{\mu}_2 \begin{bmatrix} \zeta^{-(k-1)} & -\zeta^{-(k-1)} \\ \zeta^{-k} & \zeta^{-k} \end{bmatrix}.$$

By applying the rules for the inversion of partitioned matrices, C_k ($k = 1, \ldots, 4$) can be computed from

$$\begin{bmatrix} C_1 \\ C_3 \end{bmatrix} = -c_0 \mathbf{A}_{11}^{-1} \mathbf{A}_{12} \mathbf{B}_{22} \begin{bmatrix} 0 \\ 1 \end{bmatrix}, \quad \begin{bmatrix} C_2 \\ C_4 \end{bmatrix} = c_0 \mathbf{B}_{22} \begin{bmatrix} 0 \\ 1 \end{bmatrix}$$

with $\mathbf{B}_{22} = [\mathbf{A}_{22} - \mathbf{A}_{21} \mathbf{A}_{11}^{-1} \mathbf{A}_{12}]^{-1}$. Performing the matrix inversions and multiplications,

$$C_1 = C_3 = -C_2 = -C_4 = \frac{c_0}{2(\bar{\mu}_1 \zeta^k - \bar{\mu}_2 \zeta^{-k})}. \tag{4.90}$$

So, using (4.90) in (4.89) we get

$$w_{k-1,k} = 0 \quad \text{and} \quad w_{kk} = \frac{c_0(\zeta^k - \zeta^{-k})}{\bar{\mu}_1 \zeta^k - \bar{\mu}_2 \zeta^{-k}}. \tag{4.91}$$

On the other hand, $\{w_{i,k-1}\}_{i=1}^k$ satisfy equations (4.84) – (4.88) with the second subscript k replaced by $k-1$ and the 0 and 1 on the left hand side of (4.86) and (4.87) interchanged. Hence, $w_{k-1,k-1}$ and $w_{k,k-1}$ can be found in a similar way as

$$w_{k-1,k-1} = w_{kk} \quad \text{and} \quad w_{k,k-1} = 0. \tag{4.92}$$

Since $\bar{\mathbf{T}}_k^{-1}$ has a double symmetry inherited from that of $\bar{\mathbf{T}}_k$ we have

$$\mathbf{S}_{2\infty} = \frac{1}{c_0} \lim_{k \to \infty} \begin{bmatrix} w_{kk} & w_{k,k-1} \\ w_{k-1,k} & w_{k-1,k-1} \end{bmatrix}. \tag{4.93}$$

In view of (4.29), $\bar{\mu}_1 > \bar{\mu}_2$ and $c_2^2 = \bar{\mu}_1 \bar{\mu}_2$. Hence, we also have $|\zeta| > 1$. Therefore, using (4.91) and (4.92) in (4.93) we finally get

$$\mathbf{S}_{2\infty} = \frac{1}{\bar{\mu}_1} \mathbf{I}_2.$$

□

Chapter 4. Sequential Analysis of AR Processes 79

Appendix 4.B Derivation of Relation (4.58)

To prove (4.58), let us denote $[y(n+1), \ldots, y(K)]^T$ by $\mathbf{y}_{K-n}$ and $[\hat{\mathbf{y}}_n^T, \mathbf{y}_{K-n}^T]^T = [y(n-n_a+1), \ldots, y(K)]^T$ by $\bar{\mathbf{y}}$. Because of the stationarity of the AR process, (4.8) can be written as

$$f(\mathbf{y}_{K-n} \mid \hat{\mathbf{y}}_n) = \frac{1}{(\sigma\sqrt{2\pi})^{K-n}} \exp\left\{-\frac{1}{2\sigma^2}\bar{\mathbf{y}}^T \mathbf{Q}^T \mathbf{Q}\bar{\mathbf{y}}\right\}$$

where Q is given by (4.7) for $k = K - n$. Hence, in analogy with (4.13),

$$\begin{aligned} M_{K-n}(t \mid \hat{\mathbf{y}}_n) &= E\{\exp[(z_{n+1} + \cdots + z_K)t] \mid \hat{\mathbf{y}}_n\} \\ &= \frac{1}{(\sigma\sqrt{2\pi})^{K-n}} \int_{-\infty}^{\infty} \exp\left\{-\frac{1}{2\sigma^2}\bar{\mathbf{y}}^T \mathbf{A}\bar{\mathbf{y}}\right\} d\mathbf{y}_{K-n} \end{aligned} \quad (4.94)$$

with $\mathbf{A}$ denoting the $(K - n + n_a) \times (K - n + n_a)$ matrix $(\mathbf{Q}_1^T \mathbf{Q}_1 - \mathbf{Q}_0^T \mathbf{Q}_0)t + \mathbf{Q}^T \mathbf{Q}$. By using the same notation as in the proof of Lemma 4.1 and omitting t in the argument, $\mathbf{A}$ can be partitioned as

$$\mathbf{A} = \begin{bmatrix} \mathbf{G}_1 - \mathbf{D} & \mathbf{X}^T & \mathbf{0} \\ \mathbf{X} & & \\ \mathbf{0} & & \mathbf{B}_{K-n} \end{bmatrix}.$$

For values of t such that $\mathbf{A}$ is positive definite, (4.94) yields

$$M_{K-n}(t \mid \hat{\mathbf{y}}_n) = \frac{\exp\left\{-\frac{1}{2\sigma^2}\hat{\mathbf{y}}_n^T[\mathbf{G}_1 - \mathbf{D} - \mathbf{L}\mathbf{Z}_{n_a,K-n}\mathbf{L}^T]\hat{\mathbf{y}}_n\right\}}{|\mathbf{B}_{K-n}|^{1/2}}. \quad (4.95)$$

Using the same arguments as in the proof of Lemma 4.1, we have

$$|\mathbf{B}_{K-n}| \sim |\mathbf{G}_2 - \mathbf{L}\mathbf{S}_{n_a\infty}\mathbf{L}^T| \, \Omega \, \mathcal{G}^{K-n-n_a} \quad (4.96)$$

for large $K - n$. Substituting (4.96) into (4.95) and considering the fact that $\mathbf{Z}_{n_a,K-n} \to \mathbf{S}_{n_a\infty}$ as $K \to \infty$, we obtain $M_{K-n}(t \mid \hat{\mathbf{y}}_n) \sim D(t, \hat{\mathbf{y}}_n) \lambda^{K-n}(t)$ where

$$D(t, \hat{\mathbf{y}}_n) = \frac{\exp\left\{-\frac{1}{2\sigma^2}\hat{\mathbf{y}}_n^T[\mathbf{G}_1 - \mathbf{D} - \mathbf{L}\mathbf{S}_{n_a\infty}\mathbf{L}^T]\hat{\mathbf{y}}_n\right\} \mathcal{G}^{n_a/2}}{|\mathbf{G}_2 - \mathbf{L}\mathbf{S}_{n_a\infty}\mathbf{L}^T|^{1/2} \, \Omega^{1/2}}.$$

□

Chapter 5

Change Detection in Dynamical Systems

The SPRT, which is most suitable for testing two hypotheses against each other, can be adapted to detecting changes which may occur at an unknown time. This adaptation leads to the cumulative sum (CUSUM) test which is discussed in this chapter.

In the first section, the change detection problem is contrasted with hypothesis testing and the CUSUM test is introduced. Also, the application of the CUSUM test to controlled autoregressive moving average (CARMA) processes is discussed. Section 2 introduces the quantities characterizing the performance of a CUSUM test. Examples are presented in Section 3. Section 4 concludes this chapter.

5.1 CUSUM Test

5.1.1 Rationale and definition

As explained in the previous chapter, SPRT can be used quite efficiently to decide in which mode a dynamical system is operating. That is, it can be employed as a tool for *model discrimination*. At this point, a distinction must be made between model discrimination and change detection problems. In the former the whole data set collected during the test is assumed to be generated by only one of the models describing the two hypotheses. In the latter, however, the observations are collected

Chapter 5. Change Detection in Dynamical Systems

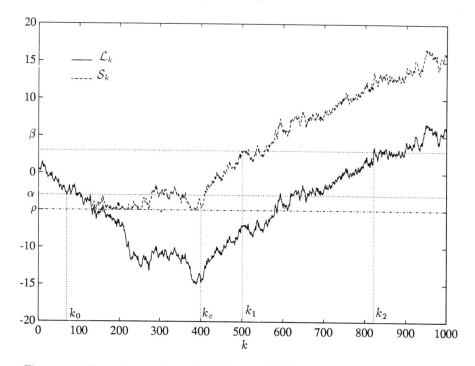

Figure 5.1: Typical behaviour of SPRT and CUSUM test in detecting a change

under $\mathcal{H}_0$ up to some unknown change time, and thereafter they are generated by the model corresponding to $\mathcal{H}_1$. In order to react to such changes as soon as possible the SPRT must be used in a modified way.

Figure 5.1 shows the typical behaviour of SPRT when a change occurs during the test. For $1 \leq k < k_c$, the system is operating under $\mathcal{H}_0$, so the log likelihood ratio, $\mathcal{L}_k$, decreases on average during this interval. It crosses the lower threshold α at k_0, which makes the statistician decide that $\mathcal{H}_0$ is true. The system changes its mode at k_c, and hence, the average trend of the log likelihood ratio is in the positive direction when $k \geq k_c$. Note that the sample number $k_2 - k_c$ required to reach the upper threshold β is quite large because of the low value of $\mathcal{L}_{k_c}$. Obviously, the later the change occurs the more time it takes to detect it.

Figure 5.1 also illustrates a way of improving the detection time, namely, resetting the log likelihood ratio back to a negative threshold ρ when it falls below it. In this case, the test statistic is computed recursively as

$$S_k = \max[\rho, S_{k-1} + z_k] \qquad k \geq 1$$

with $S_0 = 0$. The change from $\mathcal{H}_0$ to $\mathcal{H}_1$ is detected with a delay of only $k_1 - k_c$ sampling instants.

In most practical cases, $\mathcal{H}_0$ represents the normal operation mode of the system being monitored, and one is interested in detecting a transition to an anomalous mode $\mathcal{H}_1$. If one has enough confidence that $\mathcal{H}_0$ is the true hypothesis at the beginning of the test, then the log likelihood ratio can be initialized to $S_0 = \rho$. Quantitatively, this means that $p_0(\theta_0) \geq 1/(1 + e^\rho)$ in view of (3.6). It follows that the test can be conducted by comparing

$$S_k \gtrless \bar{\beta} \tag{5.1}$$

where $\bar{\beta} = \beta - \rho$ and redefining the cumulative sum S_k as

$$S_k = \max[0, S_{k-1} + z_k] \qquad \text{for} \quad k \geq 1 \tag{5.2}$$

with $S_0 = 0$. An alarm is activated if $S_k \geq \bar{\beta}$.

The cumulative sum test described by (5.1) and (5.2) was originally proposed by Page (1954). It has been used by Nikiforov (1979, 1986) to detect changes in the parameters, as well as the mean of the output, of ARMA processes. Zhang (1989) considered also a two sided resetting mechanism for the cases where changes in either direction are of interest, which had been discussed also by Page (1954).

We note from (5.2) and (3.2) that S_k accumulates the increments since that sampling instant when the log likelihood ratio has reached its minimum, i.e.,

$$\begin{aligned} S_k &= \mathcal{L}_k - \min_{1 \leq i \leq k} \mathcal{L}_i \\ &= \max_{1 \leq i \leq k} \sum_{j=i}^{k} z_j \end{aligned} \tag{5.3}$$

On the other hand, when a change is detected, the maximum likelihood estimate for the change time can be found by (Basseville, 1986)

$$\begin{aligned}\hat{k}_c &= \arg\max_{1 \leq k_c \leq k} f(\mathbf{y}_k \mid k_c) \\ &= \arg\max_{1 \leq k_c \leq k} f_0(\mathbf{y}_{k_c-1}) f_1(y(k_c), \ldots, y(k) \mid \mathbf{y}_{k_c-1}).\end{aligned}$$

The expression inside the arg max operation can be divided by $f_0(\mathbf{y}_k)$ which is independent of k_c. So,

$$\begin{aligned}\hat{k}_c &= \arg\max_{1 \leq k_c \leq k} \frac{f_1(y(k_c), \ldots, y(k) \mid \mathbf{y}_{k_c-1})}{f_0(y(k_c), \ldots, y(k) \mid \mathbf{y}_{k_c-1})} \\ &= \arg\max_{1 \leq k_c \leq k} \sum_{j=k_c}^{k} z_j. \end{aligned} \quad (5.4)$$

A comparison of (5.4) with (5.3) reveals that the maximum likelihood estimate for the change time is the instant immediately after the last resetting.

5.1.2 Application to CARMA models

Let us now consider the CUSUM test for detecting a change in the dynamics of a CARMA process. The observations are therefore generated by

$$A(q^{-1}) y(k) = q^{-d} B(q^{-1}) u(k) + C(q^{-1}) \epsilon(k). \quad (5.5)$$

where $u(k)$ is an auxiliary input. We assume that $A(z^{-1})$ and $C(z^{-1})$ polynomials are monic and have all their zeros inside the unit circle, and $d > 0$. Further, the white Gaussian noise $\epsilon(k)$ is assumed to be of zero mean and variance σ^2.

We are interested in detecting a change from $\mathcal{H}_0$ to $\mathcal{H}_1$ which are two hypotheses concerning

$$\boldsymbol{\theta} = [a_1, \ldots, a_{n_a}, b_0, \ldots, b_{n_b}, c_1, \ldots, c_{n_c}]^T,$$

namely, the coefficients of the $A(q^{-1})$, $B(q^{-1})$ and $C(q^{-1})$ polynomials. The $A(q^{-1})$, $B(q^{-1})$ and $C(q^{-1})$ polynomials corresponding to the i-th hypothesis will be denoted by $A_i(q^{-1})$, $B_i(q^{-1})$ and $C_i(q^{-1})$, respectively. Note that the noise variance σ^2 is the same under both hypotheses.

The increments of the cumulative sum are obtained in a similar way to (4.6):

$$z_k = \ln \frac{f_1(y(k) \mid \varphi(k-1))}{f_0(y(k) \mid \varphi(k-1))} \tag{5.6}$$

where $\varphi(k-1) = [y(k-1), \ldots, y(k-n_a), u(k-d), \ldots, u(k-d-n_b)]^T$. The conditional density of $y(k)$ is given by

$$f_i(y(k) \mid \varphi(k-1)) = \frac{1}{\sqrt{2\pi}\sigma} \exp\{-\frac{1}{2\sigma^2} e_i^2(k)\} \qquad i = 0, 1 \tag{5.7}$$

where

$$e_i(k) = y(k) - \hat{y}_i(k) \tag{5.8}$$

is the prediction error of the one-step-ahead output predictor based on the hypothesis $\mathcal{H}_i$. The one-step-ahead predictions are obtained from (Wellstead and Zarrop, 1991)

$$\hat{y}_i(k) = \frac{1}{C_i(q^{-1})} \left[(C_i(q^{-1}) - A_i(q^{-1})) y(k) + q^{-d} B_i(q^{-1}) u(k) \right], \qquad i = 0, 1. \tag{5.9}$$

So, from (5.6) and (5.7),

$$z_k = \frac{1}{2\sigma^2} \left[e_0^2(k) - e_1^2(k) \right]. \tag{5.10}$$

That is, S_k accumulates the information about the systems being under $\mathcal{H}_0$ or $\mathcal{H}_1$ in terms of the prediction errors corresponding to these hypotheses. In fact, $|E\{z_k\}|$ can be viewed as a *distance* between two models and is sometimes called the *Kullback divergence* (Kullback, 1959).

The prediction error based on $\mathcal{H}_i$ can be written from (5.8) and (5.9) as

$$e_i(k) = \frac{1}{C_i(q^{-1})} \left[A_i(q^{-1}) y(k) - q^{-d} B_i(q^{-1}) u(k) \right], \qquad i = 0, 1. \tag{5.11}$$

When the system operates under $\mathcal{H}_1$, by substituting $y(k)$ from (5.5) with $i = 0$ or 1 into (5.11), one finds

$$e_0(k) = \frac{A_0(q^{-1}) C_1(q^{-1})}{A_1(q^{-1}) C_0(q^{-1})} \epsilon(k) + \frac{q^{-d} \left(A_0(q^{-1}) B_1(q^{-1}) - A_1(q^{-1}) B_0(q^{-1}) \right)}{A_1(q^{-1}) C_0(q^{-1})} u(k). \tag{5.12}$$

and
$$e_1(k) = \epsilon(k), \tag{5.13}$$

respectively. Similarly, under $\mathcal{H}_0$ one has

$$e_0(k) = \epsilon(k), \tag{5.14}$$

and

$$e_1(k) = \frac{A_1(q^{-1})\,C_0(q^{-1})}{A_0(q^{-1})\,C_1(q^{-1})}\epsilon(k) + \frac{q^{-d}\left(A_1(q^{-1})\,B_0(q^{-1}) - A_0(q^{-1})\,B_1(q^{-1})\right)}{A_0(q^{-1})\,C_1(q^{-1})}u(k). \tag{5.15}$$

It is easily seen from (5.12) that, since $A_i(q^{-1})$ and $C_i(q^{-1})$ are monic ($i = 0,1$), under any causal input law (i.e., if $u(k)$ is statistically independent of $\epsilon(l)$ for $l > k$), we have $E\{e_0^2(k) \mid \mathcal{H}_1\} > E\{\epsilon^2(k)\} = \sigma^2$. On the other hand, by (5.13), $E\{e_1^2(k) \mid \mathcal{H}_1\} = \sigma^2$. Similarly, $E\{e_0^2(k) \mid \mathcal{H}_0\} = \sigma^2$ and $E\{e_1^2(k) \mid \mathcal{H}_0\} > \sigma^2$. Therefore, in view of (5.10)

$$E\{z_k \mid \mathcal{H}_0\} < 0 \qquad \text{and} \qquad E\{z_k \mid \mathcal{H}_1\} > 0. \tag{5.16}$$

So, on average, the trend of the cumulative sum is in the positive direction when $\mathcal{H}_1$ is true. However, when $\mathcal{H}_0$ is the true hypothesis, the sum tends to stick to the resetting threshold.

5.2 Performance Measures of CUSUM Test

A distinguishing feature of the change detection problem, in contrast to standard hypothesis testing, is that the test is terminated only if a decision is made in favour of $\mathcal{H}_1$. Therefore, the false alarm probability (probability of declaring a transition from $\mathcal{H}_0$ to $\mathcal{H}_1$ when $\mathcal{H}_0$ has always been true) is unity, and the missed alarm probability is zero. Hence, these quantities lose their meanings as performance measures of the test.

In fact, the aim of a change detection mechanism must be to react to the change as quickly as possible *if* it occurs, while having reasonably long times between false alarms if there is no change. Therefore, the performance of a CUSUM test is suitably characterized by its *average detection delay* and *mean time between false alarms* (Basseville, 1988).

The performance of the CUSUM test has been investigated in detail from this point of view by Lorden (1971), Shiryaev (1978) and more recently by Moustakides (1986) for the case where the samples are i.i.d. under both hypotheses. Their results indicate the following optimality property of the CUSUM test for the i.i.d. case: It minimizes the average detection delay for a fixed false alarm rate.

These two quantities can be obtained by evaluating the *average run length* (ARL) under different hypotheses. If we denote the run length of the test as $\bar{n}$ the mean time for false alarm will be $E\{\bar{n} \mid \mathcal{H}_0\}$. On the other hand, if one assumes that $S_k \approx 0$ at the time the change occurs, the average detection delay will be $E\{\bar{n} \mid \mathcal{H}_1\}$. In fact Benveniste (1986) proved that $E\{\bar{n} \mid \mathcal{H}_1\}$ is an upper bound for the conditional average detection delay, namely,

$$E\{\bar{n} \mid \mathcal{H}_1\} \geq \operatorname{ess\,sup} E\{k_d - k_c + 1 \mid y(1), \ldots, y(k_c - 1)\} \qquad k_d > k_c \qquad (5.17)$$

where k_d and k_c are the detection and onset time of the change, respectively. Equation (5.17) is a formalization of the intuitive idea that the worst possible detection delay is obtained if the cumulative sum is exactly at the resetting threshold when the change occurs.

In order to derive an expression for the ARL, we will follow a limiting approach proposed by Reynolds (1975) and Nikiforov (1979). Although Nikiforov (1979) considered an i.i.d. case, his results regarding the ARL can be restated for the dynamic case thanks to the parallelism between the ASN and OC formulae of the SPRT in the i.i.d. case on one hand and the dynamic case on the other, which was shown in Chapter 4.

Let us first note that the CUSUM test described in (5.1) and (5.2) is equivalent to the following procedure:

1. Conduct an SPRT with thresholds 0 and $\bar{\beta}$.

2. If the decision is in favour of $\mathcal{H}_0$, go to 1.

   Otherwise, stop and declare a change.

So, one would have made a number of decisions in favour of $\mathcal{H}_0$ before the first decision for $\mathcal{H}_1$. The average number of such decisions is $P_0/(1-P_0)$ (Page, 1954), with P_0 being the probability of terminating the SPRT test at the lower (zero) threshold. Obviously, the run length of the CUSUM test is the sum of the sample numbers of the SPRT's deciding for $\mathcal{H}_0$ plus the sample number of the SPRT which decides in favour of $\mathcal{H}_1$ to terminate the test procedure. Therefore,

$$E\{\bar{n}\} = \frac{P_0}{1-P_0} E\{n \mid \mathcal{L}_n < 0\} + E\{n \mid \mathcal{L}_n > \bar{\beta}\}$$
$$= \frac{E\{n\}}{1-P_0}.$$

Hence, the ARL can be determined via the ASN and OC function of an SPRT with thresholds 0 and $\bar{\beta}$. The approximations for these quantities break down, however, for $\alpha = 0$. In this case (4.68) and (4.70) yield $E\{n\} \approx 0$ and $P_\alpha \approx 1$. However, an approximate expression for the ARL can be obtained by first substituting the ASN and OC of the SPRT between $-\varepsilon$ ($\varepsilon > 0$) and $\bar{\beta}$ and then letting $\varepsilon \to 0$ (Nikiforov, 1979):

$$E\{\bar{n}\} = \lim_{\varepsilon \to 0} \frac{E\{n \mid \varepsilon\}}{1 - P_{-\varepsilon}}$$
$$\approx \lim_{\varepsilon \to 0} \frac{1}{E\{z_k\}} \left[\bar{\beta} \frac{e^{-\varepsilon t_1} - 1}{e^{-\varepsilon t_1} - e^{\bar{\beta} t_1}} - \varepsilon \frac{1 - e^{\bar{\beta} t_1}}{e^{-\varepsilon t_1} - e^{\bar{\beta} t_1}} \right] \left(\frac{e^{-\varepsilon t_1} - 1}{e^{-\varepsilon t_1} - e^{\bar{\beta} t_1}} \right)^{-1}.$$

Rearranging and taking the limit, we find

$$E\{\bar{n}\} \approx \frac{1 + \bar{\beta} t_1 - e^{\bar{\beta} t_1}}{E\{z_k\} t_1} \tag{5.18}$$

for $E\{z_k\} \neq 0$. For the case $E\{z_k\} = 0$, a similar calculation using the ASN and OC formulae in (4.69) and (4.71) yields $E\{\bar{n}\} = \bar{\beta}^2 / E\{z_k^2\}$.

Since $t_1 = 1(-1)$ if $\mathcal{H}_0(\mathcal{H}_1)$ is true, the ARL's under either hypothesis turn out to be

$$E\{\bar{n} \mid \mathcal{H}_0\} \approx \frac{\bar{\beta} + 1 - e^{\bar{\beta}}}{E\{z_k \mid \mathcal{H}_0\}} \tag{5.19}$$

and
$$E\{\bar{n} \mid \mathcal{H}_1\} \approx \frac{\bar{\beta} - 1 + e^{-\bar{\beta}}}{E\{z_k \mid \mathcal{H}_1\}}. \tag{5.20}$$

Let us note the contrast between (5.20) and the approximation proposed by Zhang (1989),
$$E\{\bar{n} \mid \mathcal{H}_1\} \approx \frac{\bar{\beta}}{E\{z_k \mid \mathcal{H}_1\}} \tag{5.21}$$
which is an intuitive expression of the sample number for the log likelihood ratio to travel from 0 to $\bar{\beta}$. Comparing (5.21) with (5.20), it is seen that (5.21) can yield a reasonable approximation only if $\bar{\beta} \gg 1$.

5.3 Examples

Example 5.1 As the first example we shall consider a first order autoregressive process as in (4.25). The average increments of the cumulative sum, $E\{z_k\}$, and t_1 have been found in Example 4.3 as a function of the hypothesized and true values of the regression parameter a_1. If the hypotheses are given as in (4.76) and the threshold of the cumulative sum test is taken to be $\bar{\beta} = 3$, then the ARL can be found using (4.78) and (4.81) in (5.18) as

$$E\{\bar{n} \mid a_{01} = 0.4, a_{11} = 0.5, a_1\} \approx \frac{(a_1^2 - 1)\left(28 - 60a_1 - e^{27}e^{-60a_1}\right)}{0.005(20a_1 - 9)^2}. \tag{5.22}$$

On the other hand, if the hypotheses are taken as
$$\begin{aligned}\mathcal{H}_0 &: a_1 = a_{01} = 0.4 \\ \mathcal{H}_1 &: a_1 = a_{11} = 0.6\end{aligned} \tag{5.23}$$
one has from (4.78) $t = 5 - 10a_1$ and from (4.81) $E\{z_k\} = 0.1(2a_1 - 1)(1 - a_1^2)$. Hence, the ARL will be

$$E\{\bar{n} \mid a_{01} = 0.4, a_{11} = 0.6, a_1\} \approx \frac{(a_1^2 - 1)\left(16 - 30a_1 - e^{15}e^{-30a_1}\right)}{0.02(10a_1 - 5)^2}. \tag{5.24}$$

The ARL's for these two cases are plotted against a_1 in Figure 5.2 where also the results of the simulations are shown. For small values of a_1 the approximations (5.22) and (5.24) seem to be less than the estimated values. Nevertheless, they

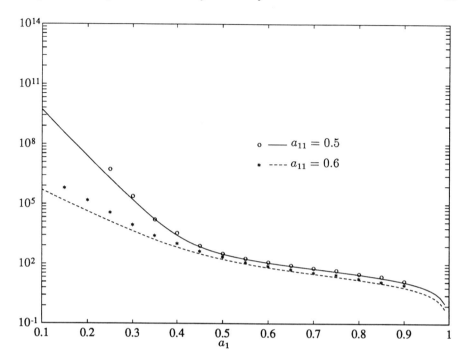

Figure 5.2: ARL of the CUSUM test in testing first order AR parameter

yielded satisfactorily close values to the estimated ones when $\mathcal{H}_0$ or $\mathcal{H}_1$ is true. Note that the ARL is a monotone decreasing function of a_1. Hence, in testing the composite hypotheses, $\mathcal{H}_0$: $a_1 < a_{01}$ and $\mathcal{H}_1$: $a_1 > a_{11}$, the false alarm rate and detection delay of the test designed for simple hypotheses will form a lower bound to the test performance.

The hypothesis pair in (5.23) are further apart from each other as compared to those in (4.76). Therefore, in the former case the average detection time is shorter than that in the latter. However, it is interesting to note from Figure 5.2 that in the case where the difference between the hypotheses is larger, the false alarm rate is higher. This point will be further discussed in the next chapter.

Example 5.2 Next we consider the following ARMA process:

$$y(k) + a_1 y(k-1) = \epsilon(k) + c_1 \epsilon(k)$$

with $|a_1| < 1$ and $|c_1| < 1$. The zero mean white noise $\epsilon(k)$ is assumed to have unit variance. We use the CUSUM test to detect a change in the parameters $\boldsymbol{\theta} = [a_1 \; c_1]^T$ from $\mathcal{H}_0$ ($\boldsymbol{\theta} = \boldsymbol{\theta}_0$) to $\mathcal{H}_1$ ($\boldsymbol{\theta} = \boldsymbol{\theta}_1$) where

$$\begin{aligned}\boldsymbol{\theta}_0 &= [a_{01} \; c_{01}]^T = [0.45 \; 0.30]^T \\ \boldsymbol{\theta}_1 &= [a_{11} \; c_{11}]^T = [0.55 \; 0.25]^T.\end{aligned} \quad (5.25)$$

The scores of the CUSUM test are computed at each sampling instant via (5.10) using the prediction errors

$$e_0(k) = \frac{A_0(q^{-1})}{C_0(q^{-1})}y(k), \qquad e_1(k) = \frac{A_1(q^{-1})}{C_1(q^{-1})}y(k).$$

When the system is operating under $\mathcal{H}_1$, the average value of z_k can be found by taking the expectation of both sides in (5.10) as

$$E\{z_k \mid \mathcal{H}_1\} = \frac{1}{2}\left[E\{e_0^2(k) \mid \mathcal{H}_1\} - 1\right].$$

where the variance of $e_0(k)$ under $\mathcal{H}_1$ can be evaluated by

$$E\{e_0^2(k) \mid \mathcal{H}_1\} = \frac{1}{2\pi}\int_{-\pi}^{\pi}\left|\frac{(1+a_{01}e^{j\omega})(1+c_{11}e^{j\omega})}{(1+a_{11}e^{j\omega})(1+c_{01}e^{j\omega})}\right|^2 d\omega.$$

So, by using the parameter values in (5.25), it turns out that $E\{z_k \mid \mathcal{H}_1\} = 1.513 \times 10^{-2}$. A similar calculation under $\mathcal{H}_0$ yields $E\{z_k \mid \mathcal{H}_0\} = -0.986 \times 10^{-2}$.

Simulations have been carried out to estimate the mean time between false alarms (MTBFA) as well as the average detection delay (ADD) when the change occurs at $k_c = 100$. The estimated values for these quantities can be compared to approximate nominal values obtained from (5.19) and (5.20) in Table 5.1. The values obtained from formula (5.21) are also shown. It is seen from Table 5.1 that (5.20) approximates the average detection detection delay far better than (5.21). Note that the estimated false alarm rate is slightly better than that obtained from the approximate formula. However, the difference between the nominal and estimated values of the average detection delay is negligible.

	$\beta = 3$		$\beta = 4$		$\beta = 5$	
	ADD	MTBFA	ADD	MTBFA	ADD	MTBFA
Nominal	135	1.63×10^3	199	5.03×10^3	265	14.4×10^3
Estimated	139	1.74×10^3	204	5.26×10^3	272	15.1×10^3
App. (5.21)	198		264		330	

Table 5.1: Performance of CUSUM test in testing ARMA parameters

5.4 Conclusions

In this chapter, we have discussed a modification of SPRT suitable for detection of changes which may occur at an unknown time. This modification led to the so called CUSUM test.

The application to CARMA models has been analyzed and the performance measures for this test have been introduced. The approximate expressions for the ARL which have been derived by Nikiforov (1979) for some i.i.d. cases are shown to be valid in the dynamic case as well. The approximation for the average detection delay thus obtained turned out to be much closer to its estimated value than the intuitive one in (5.21).

In view of (5.9), (5.8) and (5.10), the average increment size (and, hence, the ARL) seem to depend on the choice of the inputs $\{u(k)\}_{k=1}^{\infty}$. The design of inputs to improve the test performance will be investigated in the next chapter.

Chapter 6

Input Design for Change Detection

The performance of statistical change detection methods can be improved by making use of inputs available for this purpose. This chapter focuses on the exploitation of such inputs.

After introducing the motivation and some related previous work in the first section, the objectives of the input design problem for change detection will be defined in Section 2. In doing this, two main issues taken into account are to improve the detection time and, at the same time, to ensure a tolerable false alarm rate. Sections 3 and 4 are devoted to the design of two different types of input, namely, offline and online inputs. The frequency domain analysis in Section 3 reveals that optimal offline input signals can be synthesized using only one or two frequencies. Section 4, on the other hand, discusses the online generation of input signals using linear output feedback. A suboptimal solution is obtained by linearizing the cost functions and constraints of the optimization problem involved. Some simulation examples are presented in Section 5. This chapter ends with some concluding remarks in Section 6.

6.1 Introduction

Although design of optimal inputs has been among the most frequently discussed issues of the system identification area, it has been rarely addressed in the context of change detection. Evidently, if one has at one's disposal an input to the system being monitored, a proper choice of that input signal should improve the performance of the detection mechanism. Naturally, such auxiliary inputs are expected not to affect the operating conditions of the system seriously. These are maintained by inputs, set points and controllers designed prior to and independent of the change detection system. Therefore, the auxiliary input signals are required to be of zero mean, and small compared to other signals within the system. Yet, we want them to improve the detection performance. The fact that one should get maximum utility from such signals stresses the relevance of the input design problem for change detection.

Some input design problems for autoregressive models in hypothesis testing context have been attacked by Uosaki and his colleagues. They have considered designing inputs under constraints on variance (Uosaki and Hatanaka, 1987) and amplitude (Uosaki, et al., 1984) and have shown that the optimal inputs maximizing the test power are composed of a finite number of frequencies in the former case and of bang-bang type in the latter. Their results were restricted to hypotheses about the order of the AR process and did not address the effect of these inputs on the false alarm performance of the test.

The only treatment of input design for change detection purposes, to the author's knowledge, is presented in the monograph by Zhang (1989), where on-line algorithms to compute inputs as well as offline input designs were investigated. She also considered applications of these inputs for fault detection and diagnosis of a chemical plant. Nevertheless, her method was based on the idea of decreasing the detection time only, and the design considerations did not include the false alarm performance. In fact, it was claimed that inputs designed to accelerate detection by increasing discrimination between $\mathcal{H}_0$ and $\mathcal{H}_1$ (in the Kullback distance sense) improve also the false alarm performance (Zhang, 1989; pp. 62–63). However, it will be clear in the forthcoming sections that the classical tradeoff in statistical hypothesis testing, namely, fast detection versus few false alarms, is still relevant from the input design point of view.

6.2 Problem Definition

If auxiliary inputs are chosen to improve the performance of the change monitoring system, they are expected to cause reduction in average detection delay, but at the same time not to degrade the false alarm rate significantly. Therefore, we adopt the following strategy in designing optimal inputs for change detection via CUSUM tests:

$$\begin{array}{ll} \underset{\mathcal{U}}{\text{minimize}} & E\{\bar{n} \mid \mathcal{H}_1\} \\ \text{subject to} & E\{\bar{n} \mid \mathcal{H}_0\} \geq K \end{array} \quad (6.1)$$

where $\mathcal{U}$ represents the admissible set of input design parameters.

We consider the detection of changes in the parameters of a CARMA model as given by (5.5). In view of (5.16), it follows from the ARL formulas (5.19) and (5.20) that the optimal input should maximize the mean value of the increments z_k under $\mathcal{H}_1$, while holding its magnitude small under $\mathcal{H}_0$. That is, the optimal input design problem can be formulated as

$$\begin{array}{ll} \underset{\mathcal{U}}{\text{maximize}} & E\{z_k \mid \mathcal{H}_1\} \\ \text{subject to} & E\{z_k \mid \mathcal{H}_0\} \geq K_z \end{array} \quad (6.2)$$

where

$$K_z = (\bar{\beta} + 1 - e^{\bar{\beta}})/K < 0, \quad (6.3)$$

with $\bar{\beta}$ being the threshold of the CUSUM test. Note that z_k for CARMA models is computed as in (5.10).

According to the constraint in (6.2), the input will delimit the trend of $\mathcal{L}_k$ in the negative direction under $\mathcal{H}_0$. Although in SPRT this will degrade the false alarm probability, in a CUSUM test it turns out to improve the mean time for false alarms.

Chapter 6. Input Design for Change Detection

To clarify this point let us assume that the hypotheses are close to each other and that one can approximate the moment generating function of z_k by using terms of its Taylor series expansion up to the second order around $tz_k = 0$,

$$E\{e^{tz_k} \mid \mathcal{H}_0\} \approx 1 + E\{z_k \mid \mathcal{H}_0\}t + \frac{1}{2}E\{z_k^2 \mid \mathcal{H}_0\}t^2. \tag{6.4}$$

On the other hand, by using (5.10) and noting that $e_0(k) = \epsilon(k)$ when $\mathcal{H}_0$ is the true hypothesis, we find that

$$\begin{aligned}
E\{e^{z_k} \mid \mathcal{H}_0\} &= \int_{-\infty}^{\infty} \exp\left\{\frac{1}{2\sigma^2}\left[e_0^2(k) - e_1^2(k)\right]\right\} \\
&\quad \times \frac{1}{\left(\sqrt{2\pi}\sigma\right)^k} \prod_{i=1}^{k} \exp\left\{-\frac{\epsilon^2(i)}{2\sigma^2}\right\} d\epsilon(1) \cdots d\epsilon(k) \\
&= \frac{1}{\left(\sqrt{2\pi}\sigma\right)^k} \int_{-\infty}^{\infty} \exp\left\{-\frac{e_1^2(k)}{2\sigma^2}\right\} \prod_{i=1}^{k-1} \exp\left\{-\frac{\epsilon^2(i)}{2\sigma^2}\right\} d\epsilon(1) \cdots d\epsilon(k).
\end{aligned} \tag{6.5}$$

It is easily seen from (5.11) and (5.5) that because of the monicity of $A_i(q^{-1})$ and $C_i(q^{-1})$ ($i = 0, 1$), under any causal input $\partial e_1(k)/\partial \epsilon(k) = 1$. Hence, it follows from (6.5) that

$$E\{e^{z_k} \mid \mathcal{H}_0\} = 1. \tag{6.6}$$

So, substituting $t = 1$ in (6.4) and using (6.6), we have

$$E\{z_k^2 \mid \mathcal{H}_0\} \approx -2E\{z_k \mid \mathcal{H}_0\}.$$

Therefore, a rough estimate for the variance of z_k under $\mathcal{H}_0$ is

$$\text{Var}\{z_k \mid \mathcal{H}_0\} \approx -E\{z_k \mid \mathcal{H}_0\}(E\{z_k \mid \mathcal{H}_0\} + 2) \tag{6.7}$$

when $E\{z_k \mid \mathcal{H}_0\} \ll 1$.

In the CUSUM test where the test statistic is not allowed to go below a fixed threshold, false alarms would result from the fluctuations of the cumulative sum above the resetting threshold. Hence, one would expect an increase in the variance of the increments to have a degrading effect on the false alarm rate. However, (6.7) implies that increasing the average magnitude of the increments under $\mathcal{H}_0$ would result in a larger variance for z_k and, hence, a shorter mean time for alarm. Recall that in Example 5.1 two pairs of hypotheses are compared and for the pair for which the hypotheses are further apart (and hence $|E\{z_k \mid \mathcal{H}_0\}|$ is larger) a shorter mean time for false alarm is obtained.

We can summarize the above discussion in the following informal statement: The input should excite the test statistics under $\mathcal{H}_1$ while trying to keep things quiet under $\mathcal{H}_0$.

Naturally, one would have additional constraints in the optimal input design problem such as constrained input or output power depending on the type of input and the set of parameters over which the optimization is made.

6.3 Offline Inputs

6.3.1 Problem refinement

In this section, we consider the design of optimal offline inputs. That is, we assume that the input $u(k)$ and the system noise $\epsilon(l)$ are statistically independent for all k and l. It is further assumed that $u(k)$ is a wide sense stationary random process so that its autocorrelations are given by

$$E\{u(k)u(k+\tau)\} = \frac{1}{2\pi} \int_{-\pi}^{\pi} e^{j\omega\tau} d\bar{\xi}(\omega).$$

The function $\bar{\xi}(\omega)$ is unique, nondecreasing, right continuous and differentiable almost everywhere in $(-\pi, \pi]$ and $\bar{\xi}(-\pi+) = 0$. It is called the *spectral distribution function* of the input. We shall find it more convenient to work with the *one-sided spectral distribution function*, which is defined as (Doob, 1953)

$$d\xi(\omega) = \begin{cases} 2d\bar{\xi}(\omega) & \text{for } \omega \in (0, \pi) \\ d\bar{\xi}(\omega) & \text{for } \omega = 0, \pi. \end{cases}$$

Note that $d\xi(\omega)$ denotes the input power in the frequency range $(\omega, \omega + d\omega)$. The region where $d\xi(\omega) \neq 0$ is called the *spectrum* of the input.

The expressions for the constraint and cost functions of the optimization problem in (6.2) can be found by taking expectations of (5.10), i.e.,

$$E\{z_k \mid \mathcal{H}_i\} = \frac{1}{2\sigma^2} \left[E\{e_0^2(k) \mid \mathcal{H}_i\} - E\{e_1^2(k) \mid \mathcal{H}_i\} \right], \qquad i = 0, 1. \tag{6.8}$$

When $\mathcal{H}_0$ is true, the prediction errors $e_0(k)$ and $e_1(k)$ are given by (5.14) and (5.15), respectively. Hence, we get

$$\begin{aligned} E\{z_k \mid \mathcal{H}_0\} &= \frac{1}{2} - \frac{1}{2\sigma^2} E\left\{ \left[\frac{q^{-d}(A_1(q^{-1})B_0(q^{-1}) - A_0(q^{-1})B_1(q^{-1}))}{A_0(q^{-1})C_1(q^{-1})} u(k) \right]^2 \right\} \\ &\quad - \frac{1}{2\sigma^2} E\left\{ \left[\frac{A_1(q^{-1})C_0(q^{-1})}{A_0(q^{-1})C_1(q^{-1})} \epsilon(k) \right]^2 \right\}, \end{aligned} \tag{6.9}$$

since $u(k)$ and $\epsilon(l)$ are independent for any k and l. Similarly, using (5.12) and (5.13) in (6.8) we have

$$\begin{aligned} E\{z_k \mid \mathcal{H}_1\} &= -\frac{1}{2} + \frac{1}{2\sigma^2} E\left\{ \left[\frac{q^{-d}(A_1(q^{-1})B_0(q^{-1}) - A_0(q^{-1})B_1(q^{-1}))}{A_1(q^{-1})C_0(q^{-1})} u(k) \right]^2 \right\} \\ &\quad + \frac{1}{2\sigma^2} E\left\{ \left[\frac{A_0(q^{-1})C_1(q^{-1})}{A_1(q^{-1})C_0(q^{-1})} \epsilon(k) \right]^2 \right\}. \end{aligned} \tag{6.10}$$

It is clear from (6.9) and (6.10) that the input which is optimal in the sense of (6.2) will

$$\text{maximize} \quad E\left\{ \left[T_1(q^{-1}) u(k) \right]^2 \right\} \tag{6.11}$$

$$\text{subject to} \quad E\left\{ \left[T_0(q^{-1}) u(k) \right]^2 \right\} \leq \bar{K} \tag{6.12}$$

where

$$T_1(q^{-1}) = \frac{A_1(q^{-1}) B_0(q^{-1}) - A_0(q^{-1}) B_1(q^{-1})}{A_1(q^{-1}) C_0(q^{-1})} \tag{6.13}$$

$$T_0(q^{-1}) = \frac{A_1(q^{-1}) B_0(q^{-1}) - A_0(q^{-1}) B_1(q^{-1})}{A_0(q^{-1}) C_1(q^{-1})} \tag{6.14}$$

and $\bar{K}$ is related to K_z via

$$\bar{K} = \sigma^2 - 2\sigma^2 K_z - E\left\{ \left[\frac{A_1(q^{-1}) C_0(q^{-1})}{A_0(q^{-1}) C_1(q^{-1})} \epsilon(k) \right]^2 \right\}. \tag{6.15}$$

Note that in view of (5.16), (6.9) and (6.10) tell us that a nonzero input causes both $E\{z_k \mid \mathcal{H}_1\}$ and $|E\{z_k \mid \mathcal{H}_0\}|$ to be larger than in the no input case.

That means, any offline input improves the average detection delay, but also deteriorates the false alarm performance.

It also implies that to have a well posed problem the bound K on the mean time for false alarms (see (6.1)) must be less than that when no input is used. One should note, however, that these assertions are justified on the grounds of the approximate ARL formulas (5.19)–(5.20) and, hence, are strictly valid only for close hypotheses.

6.3.2 Power constrained inputs

It is clear from (6.12) that, due to the bound on the false alarm rate, the optimal input will be of bounded power. Nevertheless, there might be cases where one would like to further delimit the input power. Therefore, we shall consider an additional constraint in the problem defined in (6.11) and (6.12), namely,

$$E\left\{u^2(k)\right\} \leq K_u \qquad (6.16)$$

where $K_u > 0$ is the maximum available input power. One might also like to limit the frequency band for the input signal. For example, the input may be required to be of zero mean, i.e. $d\xi(0) = 0$. We shall denote the frequency region where $d\xi(\omega)$ is allowed to be nonzero by Ω.

Let us denote by $\mathcal{P}_\Omega$ the vector space of piecewise right continuous functions on $\Omega \subseteq [0, \pi]$ and by $\mathcal{F}_\Omega \subset \mathcal{P}_\Omega$ the set of nonnegative and nondecreasing functions which are differentiable almost everywhere on Ω. Then, following Hannan (1970; Section 2.4), the optimal input design problem in (6.11) and (6.12), considered together with (6.16), can be cast in a frequency domain framework as:

$$\underset{\xi(\omega)\in\mathcal{F}_\Omega}{\text{maximize}} \quad \int_\Omega \left|T_1(e^{j\omega})\right|^2 d\xi(\omega) \qquad (6.17)$$

$$\text{subject to} \quad \int_\Omega d\xi(\omega) \leq \pi K_u \qquad (6.18)$$

$$\text{and} \quad \int_\Omega \left|T_0(e^{j\omega})\right|^2 d\xi(\omega) \leq \pi \bar{K}. \qquad (6.19)$$

This optimization problem over the function set $\mathcal{F}_\Omega$ involves constraints and a cost function which are linear in $\xi(\omega)$. As we shall see below, the set of functions

Chapter 6. Input Design for Change Detection   99

satisfying such constraints is convex. By analogy with the linear programming problem defined in $\mathbb{R}^n$ (see e.g. Shapiro, 1979), the extreme points of the set defined by the constraints play a major role in characterizing the solution to such linear optimization problems defined in function spaces (Hallenbeck and Macgregor, 1984).

The lemma below specifies the extreme points we are interested in.

Lemma 6.1 *Let*

$$\mathcal{C}_i = \left\{ \xi(\omega) \mid \xi(\omega) \in \mathcal{F}_\Omega \quad \text{and} \quad \int_\Omega h_i(\omega) \, d\xi(\omega) \leq 1 \right\} \quad i = 1, \ldots, M$$

where $h_i(\omega)$ is continuous over Ω. Then $\mathcal{D} = \bigcap_{i=1}^M \mathcal{C}_i$ is a convex subset of $\mathcal{P}_\Omega$. If $\xi_e(\omega)$ is an extreme point of $\mathcal{D}$, then its derivative vanishes almost everywhere and the number of discontinuities of $\xi_e(\omega)$ is not more than the number of indices i for which $\int_\Omega h_i(\omega) \, d\xi(\omega) = 1$.

The proof of this lemma is given in Appendix 6.A.

The next lemma states that a cost function of the type (6.17) is maximized at one of these extreme points of $\mathcal{D}$.

Lemma 6.2 *Let $\mathcal{D}$ be as in Lemma 6.1 and*

$$J(\xi(\omega)) = \int_\Omega g(\omega) \, d\xi(\omega).$$

The maximum value of $J(\xi(\omega))$ over $\mathcal{D}$ is attained at an extreme point of $\mathcal{D}$.

The proof is in Appendix 6.A.

Remark. Lemma 6.2 has a well-known analogue in linear programming problems, which states that every supporting hyperplane to a closed, bounded set in $\mathbb{R}^n$ contains at least one of its extreme points (Walsh, 1971).

Using the two lemmas above we reach the following result.

Theorem 6.1 *There exist optimal offline inputs for the problem defined by (6.17)–(6.19) whose spectra consist of at most two frequencies. If an optimal input satisfies one of the constraints in (6.18) or (6.19) with strict inequality, then its spectrum has a single frequency.*

Chapter 6. Input Design for Change Detection

Proof. The theorem follows directly from Lemmas 6.1 and 6.2 by identifying $g(\omega) = |T_1(e^{j\omega})|^2$ in Lemma 6.2 and $M = 2$, $h_1(\omega) = (\pi K_u)^{-1}$ and $h_2(\omega) = |T_0(e^{j\omega})|^2/(\pi \bar{K})$ in Lemma 6.1. □

Theorem 6.1 drastically reduces the search domain for the optimal design. To find the optimal input signal one has to consider three cases. Namely, when i) only the power constraint is active, ii) only the false alarm constraint is active, or iii) both are active, i.e., satisfied with equality.

i) First, we relax the constraint related to the false alarm rate, (6.19), and search the single frequency spectrum achieving (6.17) and satisfying (6.18) with equality. Clearly, the maximum is obtained by

$$\xi(\omega) = \pi K_u \, \mathsf{I}(\omega - \omega_1^*) \tag{6.20}$$

where $\mathsf{I}(\omega)$ is the right continuous unit step function

$$\mathsf{I}(\omega) = \begin{cases} 0 & \text{for } \omega < 0 \\ 1 & \text{for } \omega \geq 0 \end{cases} \tag{6.21}$$

and

$$\omega_1^* = \arg\max_{\omega \in \Omega} \left|T_1(e^{j\omega})\right|. \tag{6.22}$$

The input defined by (6.20) is optimal under both constraints if it satisfies (6.19) with strict inequality as well, i.e., if

$$\left|T_0(e^{j\omega_1^*})\right|^2 < \frac{\bar{K}}{K_u} \tag{6.23}$$

The optimal frequency in (6.22) is the same as that derived by Zhang (1989) to minimize the detection time subject to an input power constraint only.

ii) Next we look for the single frequency optimal design satisfying (6.19) with equality. It follows from (6.17) and (6.19),

$$\xi(\omega) = \frac{\pi \bar{K}}{|T_0(e^{j\omega_2^*})|^2} \mathsf{I}(\omega - \omega_2^*) \tag{6.24}$$

and

$$\omega_2^* = \arg\max_{\omega \in \Omega} \left|\frac{T_1(e^{j\omega})}{T_0(e^{j\omega})}\right| = \arg\max_{\omega \in \Omega} \left|\frac{A_0(e^{j\omega})C_1(e^{j\omega})}{A_1(e^{j\omega})C_0(e^{j\omega})}\right|. \tag{6.25}$$

Chapter 6. Input Design for Change Detection

To satisfy the input power constraint (6.18), we need

$$\left|T_0(e^{j\omega_2^*})\right|^2 > \frac{\bar{K}}{K_u}. \tag{6.26}$$

It follows from (6.22)

$$\left|T_1(e^{j\omega_2^*})\right| \leq \left|T_1(e^{j\omega_1^*})\right|$$

and from (6.25)

$$\left|\frac{T_1(e^{j\omega_1^*})}{T_0(e^{j\omega_1^*})}\right| \leq \left|\frac{T_1(e^{j\omega_2^*})}{T_0(e^{j\omega_2^*})}\right|$$

and, hence,

$$\left|T_0(e^{j\omega_2^*})\right| \leq \left|T_0(e^{j\omega_1^*})\right|.$$

Therefore, if either one of (6.23) and (6.26) holds with strict inequality, the other one cannot hold.

iii) Lastly, we have to consider the case where both constraints hold with equality and find a two frequency input design

$$\xi(\omega) = \pi x_1 I(\omega - \omega_1) + \pi x_2 I(\omega - \omega_2) \tag{6.27}$$

maximizing the cost function in (6.17). Using (6.27) and considering equality in both (6.18) and (6.19), one gets

$$\begin{aligned} x_1 + x_2 &= K_u \\ |T_0(e^{j\omega_1})|^2 x_1 + |T_0(e^{j\omega_2})|^2 x_2 &= \bar{K}. \end{aligned} \tag{6.28}$$

Note that, if $|T_0(e^{j\omega_1})| = |T_0(e^{j\omega_2})|$, then a nonnegative solution for x_1 and x_2 exists only if $|T_0(e^{j\omega_1})|^2 = \bar{K}/K_u$. However, in that case, x_2 (say) could be chosen to be zero and the power spectral distribution is determined with $x_1 = K_u$ and $x_2 = 0$. Hence, without loss of generality, we assume $|T_0(e^{j\omega_1})| < |T_0(e^{j\omega_2})|$. In this case, a unique positive solution to (6.28) exists only if

$$\left|T_0(e^{j\omega_1})\right|^2 \leq \frac{\bar{K}}{K_u} < \left|T_0(e^{j\omega_2})\right|^2 \tag{6.29}$$

and is given by

$$x_1(\omega_1, \omega_2) = \frac{|T_0(e^{j\omega_2})|^2 K_u - \bar{K}}{|T_0(e^{j\omega_2})|^2 - |T_0(e^{j\omega_1})|^2} \tag{6.30}$$

$$x_2(\omega_1, \omega_2) = \frac{\bar{K} - |T_0(e^{j\omega_1})|^2 K_u}{|T_0(e^{j\omega_2})|^2 - |T_0(e^{j\omega_1})|^2}. \tag{6.31}$$

Note that if neither (6.23) nor (6.26) holds then the existence of ω_1 and ω_2 satisfying (6.29) is guaranteed. By letting

$$\Omega_1 = \left\{ \omega \mid \omega \in \Omega \quad \text{and} \quad |T_0(e^{j\omega})|^2 \leq \frac{\bar{K}}{K_u} \right\} \tag{6.32}$$

$$\Omega_2 = \left\{ \omega \mid \omega \in \Omega \quad \text{and} \quad |T_0(e^{j\omega})|^2 > \frac{\bar{K}}{K_u} \right\} \tag{6.33}$$

the optimal frequencies can be found by a two-dimensional search as

$$(\omega_1^*, \omega_2^*) = \underset{\omega_1 \in \Omega_1, \omega_2 \in \Omega_2}{\arg\max} \left\{ x_1(\omega_1, \omega_2) |T_1(e^{j\omega_1})|^2 + x_2(\omega_1, \omega_2) |T_1(e^{j\omega_2})|^2 \right\}. \tag{6.34}$$

To clarify the analysis above we have the following example.

Example 6.1 Let us consider the CARMA model in (5.5) and let $\sigma^2 = 1$. Assume that $C(q^{-1}) = 1$ and the hypotheses involve a decision involving both the $A(q^{-1})$ and $B(q^{-1})$ polynomials:

$$\begin{aligned}\mathcal{H}_0: \ & A(q^{-1}) = A_0(q^{-1}) = 1 - 0.9q^{-1} + 0.2q^{-2} & B(q^{-1}) = B_0(q^{-1}) = 0.5 - 0.3q^{-1} \\ \mathcal{H}_1: \ & A(q^{-1}) = A_1(q^{-1}) = 1 - q^{-1} + 0.34q^{-2} & B(q^{-1}) = B_1(q^{-1}) = 0.4 - 0.3q^{-1}\end{aligned} \tag{6.35}$$

We also assume that the threshold of the CUSUM test is $\bar{\beta} = 5$ and there is no restriction on the spectrum of the input, i.e., $\Omega = [0, \pi]$. Evaluating (6.13) and (6.14) at $e^{j\omega}$, we find that

$$|T_1(e^{j\omega})|^2 = \frac{0.0100 - 0.0465 \cos\omega + 0.0715 \cos^2\omega - 0.0336 \cos^3\omega}{1.436 - 2.680 \cos\omega + 1.360 \cos^2\omega}$$

and

$$|T_0(e^{j\omega})|^2 = \frac{0.0100 - 0.0465 \cos\omega + 0.0715 \cos^2\omega - 0.0336 \cos^3\omega}{1.450 - 2.160 \cos\omega + 0.800 \cos^2\omega} \tag{6.36}$$

which are also shown in Figure 6.1.

The frequency which maximizes $|T_1(e^{j\omega})|$ turns out to be $\omega_1^* = \pi$ whereas $\omega_2^* = 0.735$ maximizes $|T_1(e^{j\omega})/T_0(e^{j\omega})|$. Let us assume that $K_u = 1$. Then by (6.23), ω_1^* is the frequency of the optimal input signal if

$$\bar{K} > |T_0(e^{j\omega_1^*})|^2 = 0.0366 \tag{6.37}$$

Chapter 6. Input Design for Change Detection

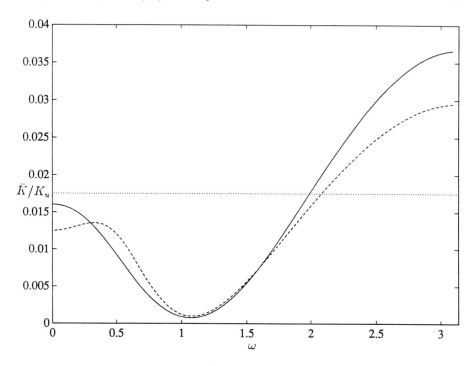

Figure 6.1: The functions $|T_0(e^{j\omega})|^2$ (solid) and $|T_1(e^{j\omega})|^2$ (dashed) in Example 6.1

To find out the corresponding bound on the mean time for false alarms one has to compute

$$E\left\{\left[\frac{A_1(q^{-1})\,C_0(q^{-1})}{A_0(q^{-1})\,C_1(q^{-1})}\epsilon(k)\right]^2\right\} = \frac{1}{\pi}\int_0^\pi \left|\frac{A_1(e^{j\omega})C_0(e^{j\omega})}{A_0(e^{j\omega})C_1(e^{j\omega})}\right|^2 d\omega.$$

By using the hypotheses in (6.35) and noting that $C_0(q^{-1}) = C_1(q^{-1}) = 1$, one finds that

$$E\left\{\left[\frac{A_1(q^{-1})\,C_0(q^{-1})}{A_0(q^{-1})\,C_1(q^{-1})}\epsilon(k)\right]^2\right\} = 1.021. \tag{6.38}$$

Using (6.38), from (6.15) and (6.3) it turns out that (6.37) is equivalent to

$$K < 4986. \tag{6.39}$$

So, as long as (6.39) holds, the optimal input is generated by

$$u(k) = \text{sign}(\phi)\cos(\pi k) \tag{6.40}$$

Chapter 6. Input Design for Change Detection 104

where ϕ is uniformly distributed on $[-\pi, \pi]$.

On the other hand, consider the case where $K > 1.161 \times 10^4$. In view of (6.3), (6.15) and (6.38), this means that $\bar{K} < 4.048 \times 10^{-3}$ which is also equal to $|T_0(e^{j\omega_2^*})|^2$. Hence, in this case the constraint on the false alarm rate dominates the power constraint (viz. (6.24)–(6.26)) and the spectral distribution of the optimal input is as in (6.24). The input signal turns out to be

$$u(k) = \frac{\sqrt{2\bar{K}}}{|T_0(e^{j\omega_2^*})|} \cos(\omega_2^* k + \phi)$$
$$= 22.1\sqrt{2\bar{K}} \cos(0.735k + \phi).$$

Finally, assume that the bound on the mean time between false alarms is such that

$$4986 < K < 1.161 \times 10^4$$

which is equivalent to the condition (6.29). In this case the spectrum of the input can consist of up to two frequencies. For example, say, $K = 7500$ which means $\bar{K} = 0.0175$. From (6.36), for $\omega = 1.991$ we have $|T_0(e^{j\omega})|^2 = \bar{K}$. From Figure 6.1 it is seen that $|T_0(e^{j\omega})|^2 \leq \bar{K}$ if $\omega \leq 1.991$. Therefore, the search regions for the optimal frequency pair are $\Omega_1 = [0, 1.991]$ and $\Omega_2 = (1.991, \pi]$. A two-dimensional numerical search yields $\omega_1^* = 0.471$ and $\omega_2^* = \pi$. Further, (6.30) and (6.31) give the power located at these two frequencies as $x_1 = 0.718$ and $x_2 = 0.282$. So, the optimal input should be generated as

$$u(k) = \sqrt{2x_1} \cos(\omega_1^* k + \phi_1) + \sqrt{x_2} \operatorname{sign}(\phi_2) \cos(\pi k)$$
$$= 1.198 \cos(0.471k + \phi_1) + 0.531 \operatorname{sign}(\phi_2) \cos(\pi k)$$

where ϕ_1 and ϕ_2 are independently and uniformly distributed in $[-\pi, \pi]$.

Some simulation results on this example are given in Section 5.

There might be cases where the optimal input has only one frequency in its spectrum even if

$$\left|T_0(e^{j\omega_2^*})\right|^2 < \frac{\bar{K}}{K_u} < \left|T_0(e^{j\omega_1^*})\right|^2$$

Chapter 6. Input Design for Change Detection

where ω_1^* and ω_2^* are as in (6.22) and (6.25), respectively. The next example illustrates such a case.

Example 6.2 As a second example consider the case where $C(q^{-1}) = 1$, $B(q^{-1}) = 0.5 - 0.3q^{-1}$ and the change occurs only in the $A(q^{-1})$ polynomial, e.g.,

$$A_0(q^{-1}) = 1 - 1.1q^{-1} + 0.3q^{-2}$$
$$A_1(q^{-1}) = 1 - q^{-1} + 0.29q^{-2}.$$

We again assume that $K_u = 1$ and $\bar{\beta} = 5$. In this case, we have

$$\left|T_0(e^{j\omega})\right|^2 = \frac{0.0101 - 0.002\cos\omega}{1.7 - 2.86\cos\omega + 1.2\cos^2\omega}$$

and

$$\left|T_1(e^{j\omega})\right|^2 = \frac{0.0101 - 0.002\cos\omega}{1.5041 - 2.58\cos\omega + 1.16\cos^2\omega}.$$

Note that $|T_1(e^{j\omega})|^2$ attains its maximum at $\omega_1^* = 0$ whereas $|T_1(e^{j\omega})/T_0(e^{j\omega})|^2$ is maximized by $\omega_2^* = 1.245$. Since $|T_0(e^{j\omega_1^*})|^2 = 0.2025$ and $|T_0(e^{j\omega_2^*})|^2 = 0.0104$, by (6.23) and (6.26), the optimal input frequency is 0 or 1.245 if either $\bar{K} > 0.2025$ or $\bar{K} < 0.0104$, respectively.

Now assume that the bound on the false alarm rate is $K = 4000$. This corresponds to $\bar{K} = 0.03868$. Therefore, a (possibly) two frequency input must be found which satisfies both constraints with equality. By (6.32) and (6.33) the frequencies must be such that $\omega_1 \in \Omega_1 = \{\omega \mid |T_0(e^{j\omega})|^2 \leq 0.03868\}$ and $\omega_2 \in \Omega_2 = \{\omega \mid |T_0(e^{j\omega})|^2 > 0.03868\}$. A two-dimensional search in this case shows that one of the maxima of $|T_1(e^{j\omega_1})|^2 x_1 + |T_1(e^{j\omega_2})|^2 x_2$ is attained if $x_2 = 0$ and $\omega_1 = \omega_2 = 0.711$. Hence, in this case, too, the optimal input consists of one frequency only.

We will return to the offline case in the simulation examples of Section 6.5.

6.4 Online Inputs

6.4.1 Problem refinement

This section focuses on the online generation of inputs to improve detection performance. We assume that the input sequence is obtained by linear output feedback

as
$$P(q^{-1})\,u(k) = F(q^{-1})\,y(k) \qquad (6.41)$$

where
$$F(q^{-1}) = f_0 + f_1 q^{-1} + \cdots + f_{n_f} q^{-n_f}$$

and
$$P(q^{-1}) = 1 + p_1 q^{-1} + \cdots + p_{n_p} q^{-n_p},$$

with n_f and n_p fixed and $P(z^{-1})$ having all its zeros inside the unit circle.

In this case, the dynamics of the system is given by

$$\left[A(q^{-1})\,P(q^{-1}) - q^{-d} B(q^{-1})\,F(q^{-1})\right] y(k) = C(q^{-1})\,P(q^{-1})\,\epsilon(k) \qquad (6.42)$$

and the problem is to determine the coefficients of the $F(q^{-1})$ and $P(q^{-1})$ polynomials so that optimal inputs in the sense of (6.2) are achieved.

It is clear from (6.42) that, although inputs of the form (6.41) do not introduce any biases into the system, they affect its poles and zeros, which might not be desirable. Therefore, one has to impose additional constraints to limit the alteration of the plant dynamics due to the input.

Naturally, one has to ensure that the system in (6.42) is stable under both hypotheses, i.e.,

$$A(z^{-1})\,P(z^{-1}) - z^{-d} B(z^{-1})\,F(z^{-1}) \neq 0 \qquad \forall |z| \geq 1, \qquad i = 0, 1.$$

However, in many cases the requirements at hand might be something more than merely having a stable system. Such conditions can be formulated by describing the permissible regions for poles and zeros under either hypothesis. Alternatively, they can be described by appropriate bounds on output variance such as

$$E\{y^2(k) \mid \mathcal{H}_i\} = E\left\{\left[\frac{C_i(q^{-1})\,P(q^{-1})}{A_i(q^{-1})\,P(q^{-1}) - q^{-d} B_i(q^{-1})\,F(q^{-1})} \epsilon(k)\right]^2\right\} \leq K_{yi} \qquad (6.43)$$

or by delimiting the input power

$$E\{u^2(k) \mid \mathcal{H}_i\} = E\left\{\left[\frac{C_i(q^{-1})\,F(q^{-1})}{A_i(q^{-1})\,P(q^{-1}) - q^{-d} B_i(q^{-1})\,F(q^{-1})} \epsilon(k)\right]^2\right\} \leq K_{ui}$$

Chapter 6. Input Design for Change Detection

for $i = 0, 1$.

To derive the mean value of z_k, the increments of the cumulative sum, under online input, we first note that (6.41) and (5.11) yield the following expression for the prediction error based on $\mathcal{H}_i$:

$$e_i(k) = \frac{A_i(q^{-1}) P(q^{-1}) - q^{-d} B_i(q^{-1}) F(q^{-1})}{C_i(q^{-1}) P(q^{-1})} y(k), \quad i = 0, 1.$$

In particular, when $\mathcal{H}_0$ is true, that is,

$$y(k) = \frac{C_0(q^{-1}) P(q^{-1})}{A_0(q^{-1}) P(q^{-1}) - q^{-d} B_0(q^{-1}) F(q^{-1})} \epsilon(k),$$

it turns out that

$$e_0(k) = \epsilon(k) \tag{6.44}$$

and

$$e_1(k) = \frac{(A_1(q^{-1}) P(q^{-1}) - q^{-d} B_1(q^{-1}) F(q^{-1})) C_0(q^{-1})}{(A_0(q^{-1}) P(q^{-1}) - q^{-d} B_0(q^{-1}) F(q^{-1})) C_1(q^{-1})} \epsilon(k). \tag{6.45}$$

Substituting (6.44) and (6.45) into (5.10) and taking expectations we get

$$E\{z_k \mid \mathcal{H}_0\} = \frac{1}{2} - \frac{1}{2\sigma^2} E\left\{\left[\frac{(A_1(q^{-1})P(q^{-1}) - q^{-d}B_1(q^{-1})F(q^{-1}))C_0(q^{-1})}{(A_0(q^{-1})P(q^{-1}) - q^{-d}B_0(q^{-1})F(q^{-1}))C_1(q^{-1})}\epsilon(k)\right]^2\right\}.$$

Using the monicity of $A_i(q^{-1})$, $C_i(q^{-1})$ and $P(q^{-1})$,

$$E\{z_k \mid \mathcal{H}_0\} = \frac{1}{2} - \frac{1}{2\sigma^2} E\left\{\left[\epsilon(k) + \left[\frac{(A_1(q^{-1}) P(q^{-1}) - q^{-d} B_1(q^{-1}) F(q^{-1}))C_0(q^{-1})}{(A_0(q^{-1}) P(q^{-1}) - q^{-d} B_0(q^{-1}) F(q^{-1}))C_1(q^{-1})} - 1\right]\epsilon(k)\right]^2\right\}$$

and rearranging, it follows that

$$E\{z_k \mid \mathcal{H}_0\} = -\frac{1}{2\sigma^2} E\left\{\left[\frac{\bar{A}(q^{-1}) P(q^{-1}) - q^{-d} \bar{B}(q^{-1}) F(q^{-1})}{(A_0(q^{-1}) P(q^{-1}) - q^{-d} B_0(q^{-1}) F(q^{-1})) C_1(q^{-1})}\epsilon(k)\right]^2\right\} \tag{6.46}$$

where

$$\bar{A}(q^{-1}) = A_1(q^{-1}) C_0(q^{-1}) - A_0(q^{-1}) C_1(q^{-1})$$

and

$$\bar{B}(q^{-1}) = B_1(q^{-1}) C_0(q^{-1}) - B_0(q^{-1}) C_1(q^{-1}).$$

On the other hand, similar manipulations yield

$$E\{z_k \mid \mathcal{H}_1\} = \frac{1}{2\sigma^2} E \left\{ \left[\frac{\bar{A}(q^{-1}) P(q^{-1}) - q^{-d} \bar{B}(q^{-1}) F(q^{-1})}{(A_1(q^{-1}) P(q^{-1}) - q^{-d} B_1(q^{-1}) F(q^{-1})) C_0(q^{-1})} \epsilon(k) \right]^2 \right\}.$$
(6.47)

In view of the above analysis the optimal input design problem for change detection under constrained output power can be restated as follows: Find $F(q^{-1})$ and $P(q^{-1})$ to

$$\text{maximize } E \left\{ \left[\frac{\bar{A}(q^{-1}) P(q^{-1}) - q^{-d} \bar{B}(q^{-1}) F(q^{-1})}{(A_1(q^{-1}) P(q^{-1}) - q^{-d} B_1(q^{-1}) F(q^{-1})) C_0(q^{-1})} \epsilon(k) \right]^2 \right\}$$

such that

$$E \left\{ \left[\frac{\bar{A}(q^{-1}) P(q^{-1}) - q^{-d} \bar{B}(q^{-1}) F(q^{-1})}{(A_0(q^{-1}) P(q^{-1}) - q^{-d} B_0(q^{-1}) F(q^{-1})) C_1(q^{-1})} \epsilon(k) \right]^2 \right\} \le -2\sigma^2 K_z,$$

$$P(z^{-1}) \ne 0 \quad \text{for} \quad |z| > 1$$

and (6.43) holds. Here K_z is given as in (6.3).

As formulated in this way, we have an optimization problem with nonlinear inequality constraints and cost function. A closed form solution to it has not been possible. Nevertheless, a suboptimal solution can be obtained by modifying the constraints in a relevant manner.

6.4.2 A suboptimal solution

For the rest of this section we shall restrict ourselves to the case $P(q^{-1}) = 1$ for simplicity. So, the design parameters are now only the coefficients of the $F(q^{-1})$ polynomial.

Let us assume that K in (6.1) is given as the mean time for false alarm when no input is applied to the system. Also assume that the bounds K_{yi} in (6.43) are taken as

$$K_{yi} = E \left\{ \left[\frac{C_i(q^{-1})}{A_i(q^{-1})} \epsilon(k) \right]^2 \right\} \quad i = 0, 1$$

Chapter 6. Input Design for Change Detection

which is the output variance under zero input. So, the design objective now becomes the reduction of the detection delay without affecting the false alarm rate and the output variance under either hypothesis. Note that this is not possible using offline design.

To achieve this objective in a suboptimal way, we shall consider linear (in $F(q^{-1})$) approximations to the mean values of z_k ((6.46) and (6.47)) and to the output variances under either hypothesis. As we aim at an input having no effect on certain quantities, this linearization will lead to some orthogonality constraints on the design parameters.

Consider a linear approximation to the nonlinear expression for $E\{z_k \mid \mathcal{H}_0\}$ given in (6.46). This is obtained by a first order Taylor series expansion around $F(q^{-1}) = 0$,

$$E\{z_k \mid \mathcal{H}_0\} \approx E\{z_k \mid \mathcal{H}_0\}|_{F(q^{-1})=0} + \sum_{i=0}^{n_f} f_i \frac{\partial}{\partial f_i} E\{z_k \mid \mathcal{H}_0\}\bigg|_{f_i=0}. \tag{6.48}$$

By differentiating (6.46) with respect to f_i and noting that $P(q^{-1}) = 1$, we get

$$\begin{aligned}\frac{\partial}{\partial f_i} E\{z_k \mid \mathcal{H}_0\} &= -\frac{1}{\sigma^2} E\left\{ \frac{\bar{A}(q^{-1}) - q^{-d}\bar{B}(q^{-1}) F(q^{-1})}{(A_0(q^{-1}) - q^{-d} B_0(q^{-1}) F(q^{-1})) C_1(q^{-1})} \epsilon(k) \right. \\ &\quad \left. \times q^{-(d+i)} \frac{(A_1(q^{-1}) B_0(q^{-1}) - A_0(q^{-1}) B_1(q^{-1})) C_0(q^{-1})}{(A_0(q^{-1}) - q^{-d} B_0(q^{-1}) F(q^{-1}))^2 C_1(q^{-1})} \epsilon(k) \right\}.\end{aligned} \tag{6.49}$$

Substituting (6.49) into (6.48) yields

$$\begin{aligned}E\{z_k \mid \mathcal{H}_0\} &\approx -\frac{1}{2\sigma^2} E\left\{ \left[G_0(q^{-1}) \epsilon(k)\right]^2 \right\} - \\ &\quad \frac{1}{\sigma^2} E\left\{ G_0(q^{-1}) \epsilon(k) \cdot H_0(q^{-1}) F(q^{-1}) \epsilon(k) \right\}\end{aligned} \tag{6.50}$$

where

$$G_0(q^{-1}) = \frac{\bar{A}(q^{-1})}{A_0(q^{-1}) C_1(q^{-1})}$$

and

$$H_0(q^{-1}) = \frac{q^{-d} (A_1(q^{-1}) B_0(q^{-1}) - A_0(q^{-1}) B_1(q^{-1})) C_0(q^{-1})}{A_0^2(q^{-1}) C_1(q^{-1})}.$$

Analogously, by differentiating (6.47), we have

$$\frac{\partial}{\partial f_i} E\{z_k \mid \mathcal{H}_1\} = \frac{1}{\sigma^2} E \left\{ \frac{\bar{A}(q^{-1}) - q^{-d}\bar{B}(q^{-1})F(q^{-1})}{(A_1(q^{-1}) - q^{-d}B_1(q^{-1})F(q^{-1}))C_0(q^{-1})} \epsilon(k) \right. $$
$$\left. \times q^{-(d+i)} \frac{(A_1(q^{-1})B_0(q^{-1}) - A_0(q^{-1})B_1(q^{-1}))C_1(q^{-1})}{(A_1(q^{-1}) - q^{-d}B_1(q^{-1})F(q^{-1}))^2 C_0(q^{-1})} \epsilon(k) \right\}$$

and, hence,

$$E\{z_k \mid \mathcal{H}_1\} \approx \frac{1}{2\sigma^2} E\left\{ \left[G_1(q^{-1})\epsilon(k) \right]^2 \right\} + \frac{1}{\sigma^2} E\left\{ G_1(q^{-1})\epsilon(k) \cdot H_1(q^{-1})F(q^{-1})\epsilon(k) \right\}$$
(6.51)

where

$$G_1(q^{-1}) = \frac{\bar{A}(q^{-1})}{A_1(q^{-1})C_0(q^{-1})}$$

and

$$H_1(q^{-1}) = \frac{q^{-d}\left(A_1(q^{-1})B_0(q^{-1}) - A_0(q^{-1})B_1(q^{-1})\right)C_1(q^{-1})}{A_1^2(q^{-1})C_0(q^{-1})}.$$

First order approximations to the output variance can be obtained via similar manipulations as

$$E\{y^2(k) \mid \mathcal{H}_i\} \approx E\left\{ \left[\frac{C_i(q^{-1})}{A_i(q^{-1})}\epsilon(k)\right]^2 \right\} +$$
$$E\left\{ \frac{C_i(q^{-1})}{A_i(q^{-1})}\epsilon(k) \cdot \frac{q^{-d}B_i(q^{-1})C_i(q^{-1})F(q^{-1})}{A_i^2(q^{-1})}\epsilon(k) \right\} \quad (6.52)$$

for $i = 0, 1$.

Further assume that $n_f \geq 4$ and the coefficients of $F(q^{-1})$ are bounded by $\mathbf{f}^T\mathbf{f} \leq K_f$, where $\mathbf{f} = [f_0, \ldots, f_{n_f}]^T$. Evidently, for K_f small enough the approximations in (6.50), (6.51) and (6.52) are valid.

A suboptimal solution for the feedback polynomial $F(q^{-1})$ can be found by solving the following optimization problem:

$$\begin{array}{c} \text{maximize} \quad \mathbf{f}^T\mathbf{s} \\ \mathbf{f} \\ \text{subject to} \quad \mathbf{f}^T\mathbf{r}_1 = \mathbf{f}^T\mathbf{r}_2 = \mathbf{f}^T\mathbf{r}_3 = 0 \quad \text{and} \quad \mathbf{f}^T\mathbf{f} \leq K_f. \end{array} \quad (6.53)$$

Here $\mathbf{s}$, $\mathbf{r}_1$, $\mathbf{r}_2$ and $\mathbf{r}_3$ are $(n_f + 1)$-dimensional vectors with elements

$$s_i = E\left\{ G_1(q^{-1})\epsilon(k) \cdot H_1(q^{-1})\epsilon(k-i+1) \right\}, \quad (6.54)$$

Chapter 6. Input Design for Change Detection

$$r_{1i} = E\left\{G_0(q^{-1})\,\epsilon(k) \cdot H_0(q^{-1})\,\epsilon(k-i+1)\right\}, \tag{6.55}$$

$$r_{2i} = E\left\{\frac{C_1(q^{-1})}{A_1(q^{-1})}\epsilon(k) \cdot \frac{q^{-d}B_1(q^{-1})\,C_1(q^{-1})}{A_1^2(q^{-1})}\epsilon(k-i+1)\right\}, \tag{6.56}$$

$$r_{3i} = E\left\{\frac{C_0(q^{-1})}{A_0(q^{-1})}\epsilon(k) \cdot \frac{q^{-d}B_0(q^{-1})\,C_0(q^{-1})}{A_0^2(q^{-1})}\epsilon(k-i+1)\right\}, \tag{6.57}$$

respectively, for $i = 1, \ldots, n_f + 1$.

The rationale behind this approach is to maximize the mean value of z_k under $\mathcal{H}_1$, while keeping the mean value of z_k under $\mathcal{H}_0$ and the output powers under $\mathcal{H}_0$ and $\mathcal{H}_1$ unchanged. The problem formulation in (6.53) is a statement of these objectives up to first order approximations.

Note that the solution of (6.53) must satisfy the constraint on $\mathbf{f}^T\mathbf{f}$ with equality. It can be found easily by the Lagrange multiplier method. We define the cost function

$$J = \mathbf{f}^T\mathbf{s} + \mathbf{f}^T\mathbf{R}\boldsymbol{\mu} + (\mathbf{f}^T\mathbf{f} - K_f)\mu_4$$

with $\mathbf{R} = [\mathbf{r}_1\ \mathbf{r}_2\ \mathbf{r}_3]$ and, $\boldsymbol{\mu} = [\mu_1\ \mu_2\ \mu_3]^T$ and μ_4 being the Lagrange multipliers. The optimal $\mathbf{f}$ vector, $\mathbf{f}^*$, is solved from the system of equations:

$$\begin{aligned}
\nabla_{\mathbf{f}} J &= \mathbf{s} + \mathbf{R}\boldsymbol{\mu} + 2\mathbf{f}^*\mu_4 &= 0\\
\nabla_{\boldsymbol{\mu}} J &= \mathbf{R}^T\mathbf{f}^* &= 0\\
\frac{\partial J}{\partial \mu_4} &= \mathbf{f}^{*T}\mathbf{f}^* - K_f &= 0.
\end{aligned}$$

After some manipulations it turns out that

$$\mathbf{f}^* = m(\mathbf{I} - \mathbf{R}(\mathbf{R}^T\mathbf{R})^{-1}\mathbf{R}^T)\mathbf{s} \tag{6.58}$$

where the scaling factor m is determined by

$$m^2 = \frac{K_f}{\mathbf{s}^T(\mathbf{I} - \mathbf{R}(\mathbf{R}^T\mathbf{R})^{-1}\mathbf{R}^T)^2\mathbf{s}}$$

and

$$\operatorname{sign}(m) = \operatorname{sign}\left(\mathbf{s}^T(\mathbf{I} - \mathbf{R}(\mathbf{R}^T\mathbf{R})^{-1}\mathbf{R}^T)\mathbf{s}\right).$$

In other words, $\mathbf{f}^*$ is chosen in the direction which is perpendicular to the gradients of $E\{z_k \mid \mathcal{H}_0\}$, $E\{y^2(k) \mid \mathcal{H}_i\}$ ($i = 0, 1$), with respect to $\mathbf{f}$ at $\mathbf{f} = \mathbf{0}$ and has the smallest angle possible with the gradient of $E\{z_k \mid \mathcal{H}_1\}$. Note that, since we have three orthogonality constraints in (6.53), it is required that $n_f > 3$.

6.5 Simulation Examples

In this section simulations will be presented to evaluate the effect of various inputs on the detection and false alarm performance of the CUSUM test.

Example 6.1 (continued) We shall consider the model from Example 6.1 and the hypotheses in (6.35). The test threshold is chosen as $\bar{\beta} = 4$.

As shown in Section 3, the optimal offline inputs can have three different spectra according to the design objectives. In order to achieve a fair comparison between different design strategies, the objectives are now chosen in such a way as to give fixed optimal input power (namely, $E\{u^2(k)\}$=1.0, 0.5 or 0.1).

Consider the case when $E\{u^2(k)\} = 1$. Then if the false alarm constraint is relaxed (or, if the bound on the mean time between false alarms is low enough), the input will be of frequency π and given as in (6.40). If, on the other hand, the input signal is designed to bound the false alarm rate only, then it will be a cosine wave of frequency 0.735, i.e., by (6.24)

$$u(k) = \frac{\sqrt{2\bar{K}}}{|T_0(e^{0.735j})|} \cos(0.735k + \phi).$$

Therefore, the input power will be unity if $\bar{K} = |T_0(e^{0.735j})|^2 = 4.048 \times 10^{-3}$. Note that, by (6.3) and (6.15), this corresponds to a bound of $K = 4.05 \times 10^3$ on $E\{\bar{n} \mid \mathcal{H}_0\}$. If the constraints are such that the input has two frequencies (namely, $\omega_1^* = 0.471$ and $\omega_2^* = \pi$) then the input power will be unity for $\bar{K} = 19.20 \times 10^{-3}$, with $x_1 = 0.654$ and $x_2 = 0.346$. This means that the optimal input signal is

$$u(k) = 1.144 \cos(0.471k + \phi_1) + 0.588 \operatorname{sign}(\phi_2) \cos(\pi k).$$

Simulations based on 500 runs are carried out with these three input spectra and with inputs having three different power levels. The system is simulated under

Chapter 6. Input Design for Change Detection

No Input	$E\{u^2(k)\}$	White Input	Optimal Offline Inputs		
			IPC	FAC	IPC+FAC
282	1.0	172	117	222	142
	0.5	219	161	244	182
	0.1	263	245	272	265

Table 6.1: Estimated average detection delays under offline inputs

No Input	$E\{u^2(k)\}$	White Input	Optimal Offline Inputs		
			IPC	FAC	IPC+FAC
6.83×10^3	1.0	3.36×10^3	2.83×10^3	5.43×10^3	4.11×10^3
	0.5	5.00×10^3	3.81×10^3	5.51×10^3	5.11×10^3
	0.1	5.84×10^3	5.38×10^3	6.20×10^3	6.14×10^3

Table 6.2: Estimated mean times between false alarms under offline inputs

white Gaussian inputs with the same range of powers. In the simulations to estimate the average detection time the system is operated under $\mathcal{H}_0$ until $k = 100$ when the change from $\mathcal{H}_0$ to $\mathcal{H}_1$ is introduced. The estimated values of the average detection delay (ADD) and the mean time between false alarms (MTBFA) are shown in Tables 6.1 and 6.2, respectively. The last three columns in these tables give the results for the optimal input signals where the input power is delimited by a constraint directly on it (IPC), or by a false alarm constraint (FAC) or by both where the bounds K_u and $\bar{K}$ are such that a two frequency design is optimal (IPC+FAC).

From Table 6.1 it is seen that the application of offline inputs (including the auxiliary white noise) reduces the detection time. In fact the larger the input power the shorter is the average detection delay. On the other hand, Table 6.2 shows that the false alarms occur more frequently if any offline input is used. Hence, Tables 6.1 and 6.2 exemplify the tradeoff between the detection delay and false alarm rate in input design.

In Table 6.1 we note that the fastest detection is obtained if the false alarm rate

Chapter 6. Input Design for Change Detection

is not taken into account in the input design (see IPC column). However, as seen in Table 6.2, such an input does not work well as far as false alarms are concerned. Even a white noise input is better in that sense. In fact, the lowest values for the mean time between false alarms among all inputs are obtained for this type of input.

On the other hand, the best false alarm performance is obtained by the inputs designed under a false alarm constraint only. (See FAC column of Table 6.2.) It is interesting to note that in this case the detection performance is worse than those of other inputs. Nevertheless, there is still some improvement in the detection delay against the no input case.

The last columns of both tables show the results corresponding to the case where the false alarm constraint is lowered so that a two frequency design is optimal. The false alarm rates and detection delays for these inputs lie between those of other two. Note that, in this case, the optimal input performs better than the white Gaussian input in both respects.

Next, we shall examine the simulation results for the suboptimal online inputs. Recall that the design objective in the online case is to keep the false alarm rate and output variances unaffected by the feedback while reducing the detection delay.

We assume that the order of the feedback polynomial is $n_f = 4$. In order to obtain the optimal $F(q^{-1})$, one has to find the $\mathbf{s}$, $\mathbf{r}_1$, $\mathbf{r}_2$ and $\mathbf{r}_3$ vectors of which the components are given by (6.54)–(6.57). They can be computed using the general formula

$$E\left\{G(q^{-1})\epsilon(k) \cdot H(q^{-1})\epsilon(k+\tau)\right\} = \frac{\sigma^2}{2\pi}\int_{-\pi}^{\pi} G(e^{j\omega})H(e^{-j\omega})e^{-j\omega\tau}\,d\omega$$

and evaluating the integral numerically. For the hypotheses in (6.35), it is found that

$$\mathbf{s} = [0.3926 \; 1.6051 \; 1.4716 \; 0.9259]^T \times 10^{-3},$$
$$\mathbf{r}_1 = [0.2876 \; 1.4838 \; 1.2779 \; 0.8533]^T \times 10^{-3},$$
$$\mathbf{r}_2 = [0.8592 \; 0.4684 \; 0.1763 \; 0.0170]^T,$$

Chapter 6. Input Design for Change Detection

	No Input	$K_f = 0.5$ $E\{u^2(k) \mid \mathcal{H}_0\} = 0.1967$ $E\{u^2(k) \mid \mathcal{H}_1\} = 0.1744$		$K_f = 0.2$ $E\{u^2(k) \mid \mathcal{H}_0\} = 0.0732$ $E\{u^2(k) \mid \mathcal{H}_1\} = 0.0668$	
		White	Online	White	Online
ADD	282	251	224	265	250
MTBFA	6.83×10^3	5.57×10^3	6.05×10^3	6.05×10^3	6.52×10^3

Table 6.3: Estimated detection and false alarm performance under online inputs

$$\mathbf{r}_3 = [1.0842 \ 0.6767 \ 0.3922 \ 0.2177]^T.$$

Using these values in (6.58), it turns out that the suboptimal output feedback is

$$u(k) = 0.0459y(k) - 0.278y(k-1) + 0.546y(k-2) - 0.351y(k-3)$$

if $K_f = 0.5$ and

$$u(k) = 0.0290y(k) - 0.175y(k-1) + 0.345y(k-2) - 0.222y(k-3)$$

if $K_f = 0.2$.

Table 6.3 tabulates the simulations with these online inputs. It also compares the performance of online inputs to that of white Gaussian inputs. In order to be able to compare the performances of the two types of input, the auxiliary white input is scaled to give the same input power as the online input. For example, for $K_f = 0.5$ we have $E\{u^2(k) \mid \mathcal{H}_0\} = 0.1967$ and $E\{u^2(k) \mid \mathcal{H}_1\} = 0.1744$. So, simulations are run using white Gaussian input signals having a variance of 0.1967 and 0.1744 to estimate the mean time between false alarms and average detection time, respectively, in each case.

As seen in Table 6.3, in the online case the improvement in the detection delay is again at the expense of some deterioration in the false alarm performance. Although the design objectives include keeping the false alarm rate unchanged, a decrease in the mean time between false alarms seems to occur due the fact that the feedback used is only suboptimal, in the sense that it takes into account only first order approximations of $E\{z_k \mid \mathcal{H}_0\}$. Nevertheless, the online input turned out to be better than the white noise signal, since it produced greater reduction in the detection delay and less increase in the false alarm rate.

6.6 Conclusions

In this chapter, inputs have been derived to improve the detection performance of CUSUM tests applied to dynamical systems. The underlying strategy has been not only to minimize the detection time, but also to bound the false alarm rate. It has been shown that there exists a tradeoff between these two performance measures in the sense that an offline input improving one of them is bound to deteriorate the other.

The optimal offline inputs have been shown to be achievable by a single cosine wave or as the sum of two cosines. The general way in which Lemma 6.1 is stated suggests that the number of frequencies in the optimal input spectrum may be increased to satisfy any other additional constraints which can be expressed by linear functionals of the power spectral distribution of the input signal.

A suboptimal solution is proposed for the online generation of the input signal by a linear output feedback. This suboptimal solution requires the order of the feedback polynomial to be larger than the number of constraints involved in the design.

The generalization of the design techniques investigated in this chapter, and of the detection mechanism itself, to the change detection problems with more than two hypotheses will be the subject of the next chapter.

Appendix 6.A Proofs of Lemmas 6.1 and 6.2

Proof of Lemma 6.1. To establish the convexity of $\mathcal{C}_i$, one has to check if

$$\xi(\omega) = \mu\, \xi_1(\omega) + (1-\mu)\xi_2(\omega) \in \mathcal{C}_i \qquad \forall \mu \in (0,1)$$

Chapter 6. Input Design for Change Detection 117

whenever $\xi_1(\omega), \xi_2(\omega) \in C_i$. Evidently, $\xi(\omega)$ is nonnegative, nondecreasing and differentiable almost everywhere on Ω as $\xi_1(\omega)$ and $\xi_2(\omega)$ are. It also satisfies

$$\begin{aligned}\int_\Omega h_i(\omega)\, d\xi(\omega) &= \mu \int_\Omega h_i(\omega)\, d\xi_1(\omega) + (1-\mu) \int_\Omega h_i(\omega)\, d\xi_2(\omega) \\ &\leq 1\end{aligned}$$

when $\xi_1(\omega), \xi_2(\omega) \in C_i$ ($i = 1, \ldots, M$). This shows that C_i is convex. Hence, so is $\mathcal{D}$, since it is an intersection of convex sets.

Next we shall show that the derivative of $\xi(\omega) \in \mathcal{D}$ must vanish almost everywhere if $\xi(\omega)$ is an extreme point of $\mathcal{D}$. We start by assuming that there is some closed nonempty interval $\bar{\Omega} = [\omega_l, \omega_h] \subset \Omega$ where

$$S(\omega) \triangleq \frac{d\xi(\omega)}{d\omega} > 0.$$

Consider a small continuous variation $\delta(\omega)$ to $S(\omega)$ such that

$$\int_{\bar{\Omega}} h_i(\omega)\, \delta(\omega)\, d\omega = 0 \qquad i = 1, \ldots, M \tag{6.59}$$

and

$$\min_{\omega \in \bar{\Omega}} \delta(\omega) \geq -\min_{\omega \in \bar{\Omega}} S(\omega). \tag{6.60}$$

A straightforward way to construct $\delta(\omega)$ is by considering an expansion of $h_i(\omega)$ and $\delta(\omega)$ into a set of orthonormal basis functions $\{\phi_k(\omega)\}_{k=1}^\infty$ over $\bar{\Omega}$,

$$h_i(\omega) = \sum_{k=1}^\infty \alpha_{ik} \phi_k(\omega) \qquad \text{and} \qquad \delta(\omega) = \sum_{k=1}^\infty \beta_k \phi_k(\omega).$$

Letting $\beta_k = 0$ for $k > M+1$, $\{\beta_k\}_{k=1}^{M+1}$ can be obtained by solving

$$\sum_{k=1}^{M+1} \alpha_{ik} \beta_k = 0 \qquad i = 1, \ldots, M \tag{6.61}$$

to satisfy (6.59). The set of M equations (6.61) in $M+1$ unknowns has infinitely many solutions and any one of them can be scaled down to further satisfy (6.60). Let

$$\xi_1(\omega) = \begin{cases} \xi(\omega) & \text{for } \omega < \omega_l \\ \xi(\omega) + \int_{\omega_l}^\omega \delta(x)\, dx & \text{for } \omega \in \bar{\Omega} \\ \xi(\omega) + \int_{\bar{\Omega}} \delta(x)\, dx & \text{for } \omega > \omega_h. \end{cases} \tag{6.62}$$

By this definition, $\xi_1(\omega)$ is nondecreasing, right continuous and, because of (6.60), nonnegative. By (6.59), $\xi_1(\omega) \in \mathcal{D}$. Let us also choose

$$\xi_2(\omega) = (1+\lambda)\xi(\omega) - \lambda\,\xi_1(\omega) \qquad \lambda > 0. \tag{6.63}$$

Since $\xi_1(\omega) = 0$ whenever $\xi(\omega) = 0$, $\xi_2(\omega)$ is nonnegative and also nondecreasing for λ small enough. From (6.63), it follows that

$$\int_\Omega h_i(\omega)\,d\xi_2(\omega) = \int_\Omega h_i(\omega)\,d\xi(\omega) + \lambda\left[\int_\Omega h_i(\omega)\,d\xi(\omega) - \int_\Omega h_i(\omega)\,d\xi_1(\omega)\right].$$

By (6.62) and (6.59) we have $\int_\Omega h_i(\omega)\,d\xi(\omega) = \int_\Omega h_i(\omega)\,d\xi_1(\omega)$. Hence,

$$\int_\Omega h_i(\omega)\,d\xi_2(\omega) = \int_\Omega h_i(\omega)\,d\xi(\omega) \leq 1$$

which means $\xi_2(\omega) \in \mathcal{D}$. We can rewrite (6.63) as

$$\xi(\omega) = \frac{\lambda}{1+\lambda}\xi_1(\omega) + \frac{1}{1+\lambda}\xi_2(\omega). \tag{6.64}$$

Therefore, $\xi(\omega)$ cannot be an extreme point of $\mathcal{D}$ unless $d\xi(\omega)/d\omega = 0$ almost everywhere.

To complete the proof we have to show that if $\xi(\omega)$ is an extreme point of $\mathcal{D}$ the number of discontinuities of $\xi(\omega)$ cannot be larger than the number constraints

$$\int_\Omega h_i(\omega)\,d\xi(\omega) \leq 1 \tag{6.65}$$

that it satisfies with equality. With a reordering of the indices, assume that $\xi(\omega)$ satisfies (6.65) with equality for $i = 1,\ldots,k$ and with strict inequality for $i = k+1,\ldots,N$. Also assume that $d\xi(\omega)/d\omega = 0$ and $\xi(\omega)$ has more than k discontinuities, i.e.,

$$\xi(\omega) = \sum_{i=1}^{l} x_i \mathsf{I}(\omega - \omega_i) \qquad l \geq k+1, \qquad x_i > 0 \;\; \forall i \qquad \omega, \omega_i \in \Omega$$

where $\mathsf{I}(x)$ is the unit step function defined in (6.21). Let[1]

$$\xi_1(\omega) = \sum_{i=1}^{k+1}(x_i + \epsilon_i)\mathsf{I}(\omega - \omega_i) + \sum_{i=k+2}^{l} x_i \mathsf{I}(\omega - \omega_i) \tag{6.66}$$

---

[1] If $l = k+1$, the second sum on the right hand side of (6.66) should be interpreted as zero.

Chapter 6. Input Design for Change Detection

where ϵ_i $(i = 1, \ldots, k+1)$ are chosen such that

$$\sum_{i=1}^{k+1} h_j(\omega_i)\epsilon_i = 0 \qquad j = 1, \ldots, k \qquad (6.67)$$

$$\max_{1 \leq i \leq k+1} \epsilon_i \leq \frac{1 - \sum_{i=1}^{l} h_j(\omega_i)\, x_i}{(k+1) \cdot \max\limits_{1 \leq i \leq k+1} h_j(\omega_i)} \qquad j = k+2, \ldots, l \qquad (6.68)$$

and

$$\epsilon_i \geq -x_i \qquad i = 1, \ldots, k+1. \qquad (6.69)$$

One can take any one of the infinitely many nontrivial solutions of k equations (6.67) in $k + 1$ unknowns and scale it down to satisfy (6.68) and (6.69). Since at least one of the ϵ_i is nonzero, then $\xi_1(\omega) \not\equiv \xi(\omega)$. By (6.69), $\xi_1(\omega)$ is nondecreasing and nonnegative and by (6.67) and (6.68) it satisfies (6.65) for $i = 1, \ldots, M$. Hence, $\xi_1(\omega) \in \mathcal{D}$. A second element of $\mathcal{D}$, $\xi_2(\omega)$, can be chosen as in (6.63) where λ can be made small enough so that $\xi_2(\omega)$ satisfies the inequality (6.65) for $i = 1, \ldots, M$. By (6.64), $\xi(\omega)$ can be written as a convex combination of $\xi_1(\omega), \xi_2(\omega) \in \mathcal{D}$ and, hence, it cannot be an extreme point of $\mathcal{D}$ unless its derivative vanishes almost everywhere and its number of discontinuities is less than or equal to k. □

Proof of Lemma 6.2. Let us denote the maximum of $J(\xi(\omega))$ attained over $\mathcal{D}$ as J^* and the set of functions in $\mathcal{D}$ achieving J^* as $\mathcal{C}^*$. $\mathcal{C}^*$ is convex, because if $\xi_1(\omega), \xi_2(\omega) \in \mathcal{C}^*$, then $\mu\xi_1(\omega) + (1-\mu)\xi_2(\omega) \in \mathcal{D}$ $(0 < \mu < 1)$ and

$$\begin{aligned} J(\mu\xi_1(\omega) + (1-\mu)\xi_2(\omega)) &= \mu \int_\Omega g(\omega)\, d\xi_1(\omega) + (1-\mu) \int_\Omega g(\omega)\, d\xi_2(\omega) \\ &= J^*. \end{aligned}$$

First, we shall show that if $\xi(\omega)$ is an extreme point of $\mathcal{C}^*$ then it is also an extreme point of $\mathcal{D}$. Let $\xi(\omega)$ be an extreme point of $\mathcal{C}^*$, but not of $\mathcal{D}$. Then there exist $\xi_1(\omega), \xi_2(\omega) \in \mathcal{D}$ and $\mu \in (0, 1)$ such that

$$\xi(\omega) = \mu\xi_1(\omega) + (1-\mu)\xi_2(\omega)$$

where either $\xi_1(\omega)$ or $\xi_2(\omega)$ (or both) is not in $\mathcal{C}^*$. However, in that case,

$$\begin{aligned} J(\xi(\omega)) &= \mu \int_\Omega g(\omega) \, d\xi_1(\omega) + (1-\mu) \int_\Omega g(\omega) \, d\xi_2(\omega) \\ &< J^* \end{aligned}$$

which contradicts the assumption that $\xi(\omega) \in \mathcal{C}^*$.

Next we need to show that $\mathcal{C}^*$ has an extreme point. Choose $\xi^*(\omega)$ such that for some $\omega_1 \in \Omega$

$$\xi^*(\omega) \leq \xi(\omega) \qquad \forall \xi(\omega) \in \mathcal{C}^*, \qquad \omega < \omega_1 \qquad (6.70)$$

and

$$\xi^*(\omega_1) < \xi(\omega_1) \qquad \forall \xi(\omega) \in \mathcal{C}^*.$$

So, $\xi^*(\omega)$ is the *unique* function satisfying

$$\xi^*(\omega) \leq \xi(\omega) \qquad \forall \xi(\omega) \in \mathcal{C}^*, \qquad \omega \leq \omega_1 \qquad (6.71)$$

Assume that $\xi^*(\omega)$ is not an extreme point of $\mathcal{C}^*$, i.e., for some $\xi_1(\omega), \xi_2(\omega) \in \mathcal{C}^*$ different from $\xi(\omega)$ and $\mu \in (0,1)$

$$\xi^*(\omega) = \mu \xi_1(\omega) + (1-\mu) \xi_2(\omega). \qquad (6.72)$$

From (6.70) and (6.72) it follows that $\xi(\omega) = \xi_1(\omega) = \xi_2(\omega)$ for $\omega < \omega_1$. On the other hand, if we evaluate (6.72) at $\omega = \omega_1$, we get

$$\begin{aligned} \xi^*(\omega_1) &= \mu \xi_1(\omega) + (1-\mu) \xi_2(\omega_1) \\ &\geq \min\{\xi_1(\omega_1), \xi_2(\omega_1)\} \end{aligned}$$

which contradicts either (6.71) or the uniqueness of $\xi^*(\omega)$. Therefore, $\xi^*(\omega)$ must be an extreme point of $\mathcal{C}^*$. □

Chapter 7
Multihypothesis Change Detection

Chapters 3–6 have been concerned with the two hypotheses case. Nevertheless, the tests and input design techniques discussed therein can be extended to cases where one has more than two alternative hypotheses about the data-generating mechanism. In this chapter such generalizations are considered.

The first section discusses multihypothesis sequential tests. It is focused mostly on a multihypothesis SPRT due to Armitage (1950). Armitage's test can be utilized to obtain change detection algorithms among more than two hypotheses. This is done in Section 2 where a new multihypothesis CUSUM test is introduced and compared to an alternative test proposed recently by Zhang (1989). Section 3 generalizes the ideas and methods for offline input design of the previous chapter to the multihypothesis case and illustrates them in an example. An extension for the online inputs is also briefly discussed. Some conclusions are drawn in Section 4.

7.1 Multihypothesis SPRT

This section will be concerned with the sequential hypothesis testing problem among M mutually exclusive hypotheses which will be denoted as $\mathcal{H}_0, \ldots, \mathcal{H}_{M-1}$.

One of the first attempts to generalize SPRT to obtain a multihypothesis sequential decision procedure is due to Sobel and Wald (1949). They considered the problem of testing three hypotheses concerning the mean of a normal distribution. Later, their ideas were generalized by Fleisher and Shwedyk (1980) to the case where one may have more than three hypotheses concerning the mean or the variance of a normal distribution. The sequential procedure proposed by Fleisher and Shwedyk, as well as the earlier three hypotheses version by Sobel and Wald, was based on the fact that in testing a parameter of a normal distribution the hypotheses can be *ordered*. That means, if a SPRT between $\mathcal{H}_{i-1}$ and $\mathcal{H}_i$ ($i = 1, \ldots, M-2$) terminates with an acceptance of $\mathcal{H}_{i-1}$ due to a particular sequence of collected data, then it is guaranteed that, with the same data, an SPRT between $\mathcal{H}_i$ and $\mathcal{H}_{i+1}$ does not accept $\mathcal{H}_{i+1}$. Hence, to decide among such ordered hypotheses $\mathcal{H}_0, \ldots, \mathcal{H}_{M-1}$, $M-1$ SPRT's can be run simultaneously between *neighbouring* hypotheses and finally a decision will be reached by eliminating the rejected hypotheses as the relevant SPRT's terminate. Both Sobel and Wald (1949) and Fleisher and Shwedyk (1980) have derived useful bounds for the ASN of such tests. Nevertheless, an ordering of the hypotheses as stated above may not be possible in more general cases such as those in which the hypotheses are described by more than one parameter.

Armitage (1950) proposed an alternative multihypothesis sequential test. His test makes use of all possible likelihood ratios among M hypotheses.

To describe Armitage's method, let us denote the log likelihood ratio between $\mathcal{H}_i$ and $\mathcal{H}_j$ by $\mathcal{L}_k(i,j)$, i.e.,

$$\mathcal{L}_k(i,j) = \ln \frac{f_i(\mathbf{y}_k)}{f_j(\mathbf{y}_k)} \qquad i,j = 0, \ldots, M-1$$

where $f_i(\mathbf{y}_k)$ denotes the probability density function of the data vector $\mathbf{y}_k$ when $\mathcal{H}_i$ is the true hypothesis. There are $M(M-1)$ such log likelihood ratios. However, note that

$$\mathcal{L}_k(i,j) = -\mathcal{L}_k(j,i). \tag{7.1}$$

Moreover, all $M(M-1)$ log likelihood ratios can be expressed in terms of $M-1$ independent ones which can be chosen in a number of ways. For example, one can use the log likelihood ratios of the hypotheses $\mathcal{H}_1, \ldots, \mathcal{H}_{M-1}$ against $\mathcal{H}_0$ to obtain others using

$$\mathcal{L}_k(i,j) = \mathcal{L}_k(i,0) - \mathcal{L}_k(j,0) \qquad i,j = 1,\ldots,M-1 \qquad (7.2)$$

In Armitage's multihypothesis SPRT, the log likelihood ratios are computed at each sampling instant until

$$\mathcal{L}_k(i,j) \geq \beta_{ij} \qquad \text{for all} \quad j = 0,\ldots,M-1 \quad \text{and} \quad j \neq i$$

for some $i \in \{0,\ldots,M-1\}$, where β_{ij} are predetermined nonnegative thresholds and, then the test is terminated with the acceptance of $\mathcal{H}_i$. In view of (7.1), this is equivalent to running $M(M-1)/2$ two-hypotheses SPRT's in parallel by monitoring the inequalities

$$-\beta_{ji} < \mathcal{L}_k(i,j) < \beta_{ij} \qquad i,j = 0,\ldots,M-1 \quad \text{and} \quad i > j \qquad (7.3)$$

until all the inequalities relevant to a particular hypothesis are violated at the same time so as to favour it. Note that the individual SPRT's defined by (7.3) may be continued even after crossing a threshold because some other SPRT's have not yet reached a decision to terminate the test. As shown by Armitage (1950), the test will terminate with probability one, since the termination of each individual SPRT is guaranteed.

As with the two hypotheses case, the thresholds β_{ij} can be used to control various error probabilities. Let us denote by ϵ_{ij} the probability of deciding in favour of $\mathcal{H}_i$ when $\mathcal{H}_j$ is true. So, ϵ_{ii} is the probability of making the correct decision under $\mathcal{H}_i$.

Using arguments similar to those Wald (1947) used in proving (3.4), Armitage (1950) has shown that the probabilities of error satisfy

$$\epsilon_{ii} \geq e^{\beta_{ij}} \epsilon_{ij} \qquad i,j = 0,\ldots,M-1, \quad i \neq j.$$

Since $\epsilon_{ii} \leq 1$, we have

$$\epsilon_{ij} \leq e^{-\beta_{ij}} \qquad i \neq j. \qquad (7.4)$$

Further, since the test will eventually terminate and, hence, $\sum_{i=0}^{M-1} \epsilon_{ij} = 1$, the probability of correct decision when $\mathcal{H}_j$ is true satisfies

$$\epsilon_{jj} \geq 1 - \sum_{i \neq j} e^{-\beta_{ij}}. \tag{7.5}$$

From (7.4) and (7.5), it is clear that by choosing sufficiently large thresholds one can obtain correct decision probabilities as high and error probabilities as low as desired.

Note that, in contrast to the test procedure proposed by Sobel and Wald (1949) and Fleisher and Shwedyk (1980), no assumptions are made about the nature of hypotheses in Armitage's multihypothesis SPRT. It is equally applicable to test hypotheses concerning the parameters describing the dynamics of data generators. The price paid for this general applicability is a serious theoretical difficulty: As pointed out by Ghosh (1970; p. 275) and still later by Wetherill and Glazebrook (1986; p. 43), we have practically no knowledge of the ASN of this test. Nevertheless, the procedure is intuitively appealing and, as we shall see in the forthcoming sections, it can be modified to obtain a multihypothesis CUSUM procedure which can accommodate arguments for input design in the multihypothesis case.

7.2 Multihypothesis CUSUM Test

From the discussion in Section 5.1, it should be clear that the multihypothesis SPRT (as with the two-hypothesis SPRT) may yield long delays in detecting changes from a nominal mode $\mathcal{H}_0$ to other alternatives $\mathcal{H}_1, \ldots, \mathcal{H}_{M-1}$. This arises because various log likelihood ratios, especially $\mathcal{L}_k(i,0)$ ($i = 1, \ldots, M-1$), will reach very low values (and hence be very distant from the alarm thresholds) when the data are generated under $\mathcal{H}_0$ for a long while. Therefore, to obtain fast detection one needs a generalization of the CUSUM procedure presented in Chapter 5.

A generalization to the multihypothesis case of the idea of resetting the log likelihood ratio to prevent it from becoming too low has been presented by Zhang

(1989). She has proposed resetting the log likelihood ratios concerning the normal operation mode $\mathcal{H}_0$ (namely $\mathcal{L}_k(i,0)$, $i = 1, \ldots, M-1$) independently from each other and to compute the others via (7.2). That is, in Zhang's procedure, starting with $\bar{S}_0(i,0) = 0$ ($i = 1, \ldots, M-1$) one computes the cumulative sums as

$$\bar{S}_k(i,0) = \max\left[0, \bar{S}_{k-1}(i,0) + z_k(i,0)\right] \quad i = 1, \ldots, M-1$$
$$\bar{S}_k(i,j) = \bar{S}_k(i,0) - \bar{S}_k(j,0) \quad i,j = 1, \ldots, M-1, \quad j \neq i. \quad (7.6)$$

Here, $z_k(i,0)$ is the increment of the log likelihood ratio between $\mathcal{H}_i$ and $\mathcal{H}_0$. An alarm is flagged for $\mathcal{H}_i$ ($i = 1, \ldots, M-1$) if

$$\bar{S}_k(i,j) \geq \bar{\beta}_{ij} \quad \text{for all} \quad j = 0, \ldots, M-1, \quad j \neq i \quad (7.7)$$

where $\bar{\beta}_{ij}$ are some positive thresholds.

We shall now consider an alternative test procedure which is a generalization of the CUSUM test in the multihypothesis case. It has already been discussed in Section 5.2 that the CUSUM test in the two hypotheses case is equivalent to applying SPRT's with zero lower threshold consecutively, until a decision is made in favour of $\mathcal{H}_1$. Adopting this approach, one can obtain a multihypothesis CUSUM test by the following procedure:

1. Conduct Armitage's multihypothesis SPRT with thresholds $\bar{\beta}_{0j} = 0$ and $\bar{\beta}_{ij} > 0$ ($i = 1, \ldots, M-1$, $j = 0, \ldots, M-1$, $j \neq i$).

2. If the decision is in favour of $\mathcal{H}_0$, go to 1.
   Otherwise stop and declare a change.

That means, at each sampling instant k, the cumulative sums are computed according to

$$\mathcal{S}_k(i,0) = \begin{cases} 0 & \text{if } \mathcal{S}_{k-1}(j,0) + z_k(j,0) \leq 0 \\ & \forall j \in \{1, \ldots, M-1\} \\ \mathcal{S}_{k-1}(i,0) + z_k(i,0) & \text{otherwise} \end{cases} \quad (7.8)$$

for $i = 1, \ldots, M - 1$ and

$$\mathcal{S}_k(i,j) = \mathcal{S}_k(i,0) - \mathcal{S}_k(j,0) \qquad i,j = 1, \ldots, M - 1 \quad j \neq i. \tag{7.9}$$

At sampling instants when no resetting is applied (i.e., when the current multihypothesis SPRT has not yet reached a decision) all cumulative sums will be updated recursively as $\mathcal{S}_k(i,j) = \mathcal{S}_{k-1}(i,j) + z_k(i,j)$ where

$$z_k(i,j) = z_k(i,0) - z_k(j,0) \qquad i,j = 1, \ldots, M - 1.$$

An alarm is set for the i-th mode if

$$\mathcal{S}_k(i,j) \geq \bar{\beta}_{ij} \qquad \text{for all} \quad j = 0, \ldots, M - 1, \quad j \neq i. \tag{7.10}$$

Note that the multihypothesis CUSUM test defined by (7.8) and (7.9) together with (7.10), as well as Zhang's procedure, reduce to the CUSUM test analyzed in Chapter 5 for $M = 2$. But the behaviour of the statistics used by either test may be quite different on some occasions. In Zhang's test the cumulative sums $\bar{\mathcal{S}}_k(i,0)$ are reset independently of each other. However, it can be seen from (7.6) and (7.7) that the detection time is affected by *all* the sums to which resetting is applied. Therefore, in some cases, the (correct or false) detection of $\mathcal{H}_i$, say, can be delayed due to the resetting of $\bar{\mathcal{S}}_k(j,0)$, $j \neq i$. In the new test procedure proposed above this is not the case, since the cumulative sums $\mathcal{S}_k(i,0)$, $i = 1, \ldots, M - 1$ are reset either all together or none at all. This point is best clarified with an example.

Example 7.1 Consider the problem of detecting a change in the mean θ of a sequence of normally and independently distributed random variables $\{y(k)\}_{k=1}^{\infty}$ having unit variance. Assume that θ can take one of three possible values θ_0, θ_1 or θ_2 where

$$\theta_2 < \theta_0 < \theta_1 \tag{7.11}$$

the nominal value being θ_0. A change is to be detected from $\mathcal{H}_0$: $\theta = \theta_0$ to either one of the hypotheses $\mathcal{H}_i$: $\theta = \theta_i$ ($i = 1, 2$). The cumulative sums are calculated

	$\beta_{10} = \beta_{12} = \beta_{20} = \beta_{21} = 4$	
	Multihyp. CUSUM test	Zhang's test
ADD	175	163
MTBFA	2.99×10^3	1.62×10^3

	$\beta_{10} = \beta_{20} = 4, \beta_{12} = \beta_{21} = 6$	
	Multihyp. CUSUM test	Zhang's test
ADD	177	264
MTBFA	3.10×10^3	12.00×10^3

Table 7.1: Estimated performances of Zhang's procedure and multihypothesis CUSUM test in testing the mean of a Gaussian random variable

using the scores

$$z_k(i,0) = (\theta_i - \theta_0)\, y(k) + \frac{1}{2}(\theta_0^2 - \theta_i^2), \tag{7.12}$$

in analogy with (3.16). From (7.12), we have

$$E\{z_k(i,0) \mid \mathcal{H}_j\} = (\theta_i - \theta_0)\theta_j + \frac{1}{2}(\theta_0^2 - \theta_i^2). \tag{7.13}$$

Figure 7.1 compares the typical behaviours of some cumulative sums in the multihypothesis CUSUM test ($\mathcal{S}_k(1,0)$, $\mathcal{S}_k(2,0)$ and $\mathcal{S}_k(1,2)$) and those in Zhang's test ($\bar{\mathcal{S}}_k(1,0)$, $\bar{\mathcal{S}}_k(2,0)$ and $\bar{\mathcal{S}}_k(1,2)$). Note that, in contrast to $\bar{\mathcal{S}}_k(i,0)$, the quantities $\mathcal{S}_k(i,0)$ can take negative values since resetting is applied only if all of them are negative. In generating the graphs in Figure 7.1, the values of θ_i are taken as $\theta_0 = 0$, $\theta_1 = 0.2$ and $\theta_2 = -0.2$ and a change is introduced from $\mathcal{H}_0$ to $\mathcal{H}_1$ at $k = 200$. The estimated values of average detection delays (ADD) and mean times between false alarms (MTBFA) based on 1000 runs are shown in Table 7.1 for two sets of thresholds.

After the change has occurred, both $\mathcal{S}_k(1,0)$ and $\bar{\mathcal{S}}_k(1,0)$ increase as new data are collected. On the other hand, by (7.13) and (7.11), $E\{z_k(2,0) \mid \mathcal{H}_1\} < 0$. Therefore, the trend of $\mathcal{S}_k(2,0)$ is in the negative direction and no resetting is applied to $\mathcal{S}_k(2,0)$ since $\mathcal{S}_k(1,0) > 0$ after the change. However, $\bar{\mathcal{S}}_k(2,0)$ is reset even after the change.

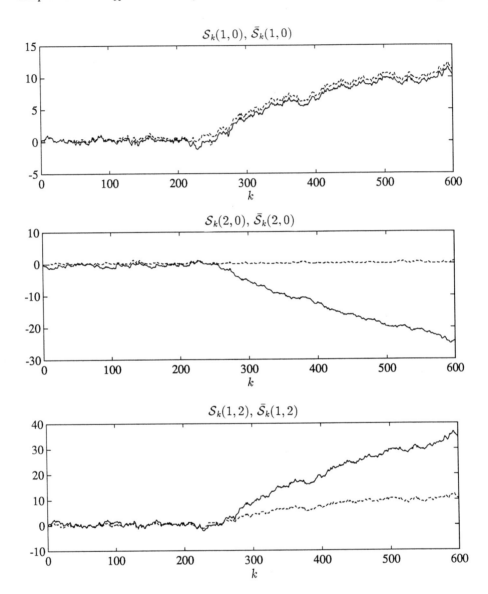

Figure 7.1: Typical behavior of the cumulative sums $\mathcal{S}_k(i,j)$ (solid lines) and $\bar{\mathcal{S}}_k(i,j)$ (dashed lines) in Example 7.1

Chapter 7. Multihypothesis Change Detection   129

Since $\mathcal{S}_k(1,2) = \mathcal{S}_k(1,0) - \mathcal{S}_k(2,0)$ and $\bar{\mathcal{S}}_k(1,2) = \bar{\mathcal{S}}_k(1,0) - \bar{\mathcal{S}}_k(2,0)$, the cumulative sum $\mathcal{S}_k(1,2)$ grows more rapidly than $\bar{\mathcal{S}}_k(1,2)$ as seen in Figure 7.1. Obviously, this is due to the resetting applied to $\bar{\mathcal{S}}_k(2,0)$ after the change.

In view of Figure 7.1, if $\bar{\beta}_{10}$ and $\bar{\beta}_{12}$ are such that the cumulative sum between $\mathcal{H}_1$ and $\mathcal{H}_0$ crosses the threshold $\bar{\beta}_{10}$ later than that between $\mathcal{H}_1$ and $\mathcal{H}_2$ crosses $\bar{\beta}_{12}$, the detection delay is typically determined by $\mathcal{S}_k(1,0)$ or $\bar{\mathcal{S}}_k(1,0)$. In that case, it seems that there is no significant difference between the detection delays in Zhang's test and the CUSUM test using $\mathcal{S}_k(i,0)$'s in (7.8). This is because, both $\bar{\mathcal{S}}_k(1,0)$ and $\mathcal{S}_k(1,0)$ are expected to be around zero before the change, but the latter might be negative. However, if $\bar{\beta}_{12}$ is considerably higher than $\bar{\beta}_{10}$, then the detection delay may be determined effectively by the cumulative sum involving $\mathcal{H}_1$ and $\mathcal{H}_2$. In this case, the multihypothesis CUSUM test which uses $\mathcal{S}_k(i,j)$'s seems to be reacting to the change more quickly than Zhang's procedure.

These effects of different resetting rules can also be observed in the simulation results presented in Table 7.1. When all the thresholds $\bar{\beta}_{ij}$ are chosen as 4, Zhang's test detects changes slightly faster than the multihypothesis CUSUM test. Nevertheless, this difference is not as big as that in the mean time for false alarms where the multihypothesis CUSUM test performs better. Also note that by choosing higher values for $\bar{\beta}_{ij}$ ($i \neq j$) (namely, $\bar{\beta}_{12} = \bar{\beta}_{21} = 6$) the performance of the multihypothesis CUSUM test is hardly affected, whereas both the average detection delay and false alarm time of the other test increase.

This example justifies the following argument for the general case: Assume that a change occurs from $\mathcal{H}_0$ to $\mathcal{H}_i$. If $E\{z_k(j,0) \mid \mathcal{H}_i\} > 0$ for all $j = 1, \ldots, M-1$ then no cumulative sum is expected to be subject to resetting after the change. So, $\mathcal{S}_k(j,0)$ and $\bar{\mathcal{S}}_k(j,0)$ behave more or less the same way. However, if $E\{z_k(j,0) \mid \mathcal{H}_i\} < 0$ for some j, then typically, $\mathcal{S}_k(j,0) < \bar{\mathcal{S}}_k(j,0)$ and, hence, $\mathcal{S}_k(i,j) > \bar{\mathcal{S}}_k(i,j)$. Therefore, the statistics $\mathcal{S}_k(i,j)$ might yield a faster detection than $\bar{\mathcal{S}}_k(i,j)$ depending on the choice of the thresholds. However, this slowness of $\bar{\mathcal{S}}_k(i,j)$ relative to $\mathcal{S}_k(i,j)$ will result in a better performance of the former in delaying the false alarm as compared to the latter.

To further improve the test performance via an auxiliary input signal, the techniques developed in the previous chapter can be generalized to the multihypothesis case. This will be the topic of the next section.

7.3 Multihypothesis Input Design

7.3.1 Problem definition

In the two hypotheses case, where we have followed the strategy of minimizing the average detection delay while bounding the mean time for a false alarm, the design of the input was eventually shown to be aimed at manipulating the average increments of the cumulative sum under $\mathcal{H}_1$ and $\mathcal{H}_0$. This method can be used in a generalized way to obtain offline input signals in the multihypothesis case.

First of all, following the derivation in Subsection 5.1.2, let us note that if the hypotheses concern the parameters of a CARMA process, the increments $z_k(i,j)$ are given by

$$z_k(i,j) = \frac{1}{2\sigma^2} \left[e_j^2(k) - e_i^2(k) \right] \qquad (7.14)$$

where $e_i(k)$ is the prediction error based on $\mathcal{H}_i$ and given by (5.8) and (5.9) where i can take values between 0 and $M-1$. Via similar manipulations used to obtain (5.12), $e_i(k)$ ($i = 0, \ldots, M-1$) will turn out to be

$$e_i(k) = \frac{A_i(q^{-1}) C_l(q^{-1})}{A_l(q^{-1}) C_i(q^{-1})} \epsilon(k) + q^{-d} T_{il}(q^{-1}) u(k), \qquad (7.15)$$

when $\mathcal{H}_l$ ($l = 0, \ldots, M-1$) is the true hypothesis, where

$$T_{il}(q^{-1}) = \frac{A_i(q^{-1}) B_l(q^{-1}) - A_l(q^{-1}) B_i(q^{-1})}{A_l(q^{-1}) C_i(q^{-1})}.$$

The subscripts on the $A(q^{-1})$, $B(q^{-1})$ and $C(q^{-1})$ polynomials are used to denote that their coefficients are to be taken as described by a particular hypothesis.

When a change occurs from $\mathcal{H}_0$ to $\mathcal{H}_i$, the cumulative sums $\mathcal{S}_k(i,j)$ ($j \neq i$) will grow on average, and the change will be detected when they reach their respective

Chapter 7. Multihypothesis Change Detection 131

thresholds. A faster detection would be expected if the auxiliary input is used in a way to increase the average increments of these cumulative sums under $\mathcal{H}_i$.

Using (7.15) in (7.14) and taking expectations, we find

$$E\{z_k(i,j) \mid \mathcal{H}_i\} = \frac{1}{2\sigma^2} E\left\{\left[\frac{A_j(q^{-1})\,C_i(q^{-1})}{A_i(q^{-1})\,C_j(q^{-1})}\epsilon(k)\right]^2\right\} - \frac{1}{2} + \frac{1}{2\sigma^2} E\left\{\left[T_{ji}(q^{-1})\,u(k)\right]^2\right\} \tag{7.16}$$

if $u(k)$ is an offline input independent of $\epsilon(k)$.

The term related to the input in (7.16) is always positive; so, *any* input will increase the mean values of *all* $z_k(i,j)$ relevant to the detection of $\mathcal{H}_i$. A reasonable choice for the input signal to get the best improvement in the detection delay is to concentrate the effect of the input on the *worst case*. By this the following is meant: Assume that all thresholds employed by the multihypothesis CUSUM test for the detection of $\mathcal{H}_i$ are equal, i.e.,

$$\bar{\beta}_{ij} = \bar{\beta}_i \qquad \forall j. \tag{7.17}$$

Then, when no input is used the cumulative sum which, on average, crosses the threshold latest, $\mathcal{S}_k(i,j_i^*)$, say, will be the one which has the smallest increment. Therefore, the input will be chosen to maximize $E\{z_k(i,j_i^*) \mid \mathcal{H}_i\}$ where, referring to (7.16), j_i^* is determined such that

$$E\left\{\left[\frac{A_{j_i^*}(q^{-1})\,C_i(q^{-1})}{A_i(q^{-1})\,C_{j_i^*}(q^{-1})}\epsilon(k)\right]^2\right\} \le E\left\{\left[\frac{A_j(q^{-1})\,C_i(q^{-1})}{A_i(q^{-1})\,C_j(q^{-1})}\epsilon(k)\right]^2\right\} \qquad \forall j = 0,\ldots,M-1. \tag{7.18}$$

By considering $E\{z_k(i,j) \mid \mathcal{H}_i\}$ as a measure of distance between $\mathcal{H}_i$ and $\mathcal{H}_j$ when $\mathcal{H}_i$ is true, the above line of thought can be interpreted as using the input to distinguish between those hypotheses which are closest to each other among all others, when no input is used.

The choice of the index j_i^* in the more general case, where (7.17) does not hold, seems to be made by ad hoc reasoning. Nevertheless, (5.20) suggests that a comparison of the quantities like $E\{z_k(i,j) \mid \mathcal{H}_i, u(k) = 0\}/(\bar{\beta}_{ij} - 1 + e^{\bar{\beta}_{ij}})$ or even simply $E\{z_k(i,j) \mid \mathcal{H}_i, u(k) = 0\}/\bar{\beta}_{ij}$ (see (5.21)) may serve as a useful guideline. From now on, for the sake of simplicity of the arguments, we shall assume that (7.17) holds so that j_i^* is chosen according to (7.18).

On the other hand, in view of the experience gained from the previous chapter, the inputs designed for accelerating detection may cause a decrease in the mean time between false alarms. Let us consider a false alarm in favour of a particular hypothesis $\mathcal{H}_i$. The run length for this false alarm to be set, under a given data sequence, in a multihypothesis CUSUM test will be larger than that necessary in a CUSUM test between $\mathcal{H}_i$ and $\mathcal{H}_0$ only, applied to the same data with the threshold $\bar{\beta}_{i0}$.

This can be clarified as follows: Consider $\mathcal{S}_k$, the statistic used in a two hypotheses CUSUM test between $\mathcal{H}_0$ and $\mathcal{H}_i$, and $\mathcal{S}_k(i,0)$, a statistic to compare $\mathcal{H}_i$ to $\mathcal{H}_0$ in the multihypothesis CUSUM test. Due to the resetting, $\mathcal{S}_k \geq 0$, whereas $\mathcal{S}_k(i,0)$ can take negative values since a resetting of $\mathcal{S}_k(i,0)$ will also depend on the values of $\mathcal{S}_k(j,0)$ ($j \neq i$) in the multihypothesis test. Since under the same data both statistics are updated using the same increments we have $\mathcal{S}_k \geq \mathcal{S}_k(i,0) \ \forall k \geq 1$. When all $\mathcal{S}_k(i,j)$ ($j = 0, \ldots, M-1, j \neq i$) cross their respective thresholds $\bar{\beta}_{ij}$, $\mathcal{S}_k$ must have already crossed $\bar{\beta}_{i0}$.

Therefore, a suitable lower bound for the mean time of a false alarm indicating the i-th mode in the multihypothesis case is the ARL of a CUSUM test between $\mathcal{H}_0$ and $\mathcal{H}_i$ when $\mathcal{H}_0$ is true. That is, in view of (5.19),

$$E\{\bar{n} \mid \mathcal{H}_0 \text{ true}, \mathcal{H}_i \text{ accepted}\} \geq \frac{\bar{\beta}_{i0} + 1 - e^{\bar{\beta}_{i0}}}{E\{z_k(i,0) \mid \mathcal{H}_0\}} \quad (7.19)$$

where $\bar{n}$ is the run length of the multihypothesis test. The mean value of $z_k(i,0)$ under $\mathcal{H}_0$ follows from (7.15) and (7.14) as

$$E\{z_k(i,0) \mid \mathcal{H}_0\} \quad (7.20)$$
$$= \frac{1}{2} - \frac{1}{2\sigma^2} E\left\{ \left[\frac{A_i(q^{-1}) C_0(q^{-1})}{A_0(q^{-1}) C_i(q^{-1})} \epsilon(k) \right]^2 \right\} - \frac{1}{2\sigma^2} E\left\{ \left[T_{i0}(q^{-1}) u(k) \right]^2 \right\}.$$

Chapter 7. Multihypothesis Change Detection

In the light of the above discussion, the problem of designing power constrained offline inputs in the multihypothesis case can be formulated as follows:

$$\underset{\mathcal{U}}{\text{maximize}} \quad \sum_{i=1}^{M-1} w_i\, E\{z_k(i, j_i^*) \mid \mathcal{H}_i\} \tag{7.21}$$

$$\text{subject to} \quad E\{z_k(i, 0) \mid \mathcal{H}_0\} \leq K_{zi} \quad i = 1, \ldots, M-1 \tag{7.22}$$

$$\text{and} \quad E\{u^2(k)\} \leq K_u. \tag{7.23}$$

Note that the cost function in (7.21) is obtained as a weighted sum of the average increments each of which is relevant to the detection of a different mode. The weights w_i can be chosen to reflect the a priori likelihood of particular changes as well as to account for the relative severity of the delays in detecting them. On the other hand, the constraints in (7.22) aim to impose lower bounds on the mean time for false alarms indicating different operation modes, via (7.19). One way to obtain the K_{zi}'s in (7.22) is to use directly (7.19) by putting the desired lower bound for the false alarm times on the left hand side. The K_{zi}'s can also be viewed as design parameters rather than requirements and can be determined by more ad hoc considerations. Clearly, (7.23) is an explicit power constraint on the input.

7.3.2 Offline inputs

Next a theorem will be presented which is an analogue of Theorem 6.1 in the multihypothesis case. Before stating it, let us write the problem expressed in (7.21)–(7.23) in the frequency domain using the notation introduced in Section 6.3. Noting in (7.16) and (7.20) that $E\{z_k(i,j) \mid \mathcal{H}_i\}$ and $E\{z_k(i,0) \mid \mathcal{H}_0\}$ are monotonically related to the second moments of the quantities $T_{ji}(q^{-1})\,u(k)$ and $T_{i0}(q^{-1})\,u(k)$, respectively, (7.21)–(7.23) is equivalent to

$$\underset{\xi(\omega) \in \mathcal{F}_\Omega}{\text{maximize}} \quad \int_\Omega \sum_{i=1}^{M-1} w_i \left|T_{j_i^* i}(e^{j\omega})\right|^2 d\xi(\omega) \tag{7.24}$$

$$\text{subject to} \quad \int_\Omega \left|T_{i0}(e^{j\omega})\right|^2 d\xi(\omega) \leq \pi \bar{K}_i \quad i = 1, \ldots, M \tag{7.25}$$

where we have defined for convenience $|T_{M0}(e^{j\omega})|^2 \equiv 1$ and $\bar{K}_M = K_u$, so that the M-th constraint corresponds to (7.23). Clearly, (7.24)–(7.25) reduces to (6.17)–(6.19) for $M = 2$.

Theorem 7.1 *There exist optimal offline inputs for the problem defined in (7.24) –(7.25) whose spectra consist of at most as many frequencies as the number of constraints satisfied by equality in (7.25).*

Proof. The theorem is a direct consequence of Lemmas 6.1 and 6.2. Namely, in these lemmas one has to identify $g(\omega) = \sum_{i=1}^{M-1} w_i \left|T_{j_i^* i}(e^{j\omega})\right|^2$ and $h_i(\omega) = |T_{i0}(e^{j\omega})|^2 / (\pi \bar{K}_i)$ to obtain the theorem. □

As some of the constraints in (7.25) may be satisfied with strict inequalities by the optimal input we can obtain the optimal design by a sequential method: One would first start with considering only one of the constraints in (7.25) and, as Theorem 7.1 implies, search for a single frequency input. If any one of such possible designs satisfies all the other constraints, that design is optimal. Otherwise, one would consider optimal designs satisfying two of the constraints and check if such an input satisfies the remaining constraints. To do this one has to consider only the case where both constraints are satisfied with equality. This is because the possibility of the existence of an optimal input satisfying only one of the constraints with equality has been discarded in the previous step of the design. By increasing the number of active constraints one will eventually reach the optimal design. The number of possible combinations of the constraints satisfied with equality is $\sum_{i=1}^{M} \binom{M}{i} = 2^M - 1$. Note that according to Theorem 7.1 at each step of the design procedure the candidate for the optimal design can be found by searching over a given number of frequencies and the input powers on each of them.

The following algorithm describes the complete design procedure:

1. Set $N_f = 1$. Let $\tilde{T}(\omega) = \sum_{i=1}^{M-1} w_i |T_{j_i^* i}(e^{j\omega})|^2$.

2. Set $N_c = \binom{M}{N_f}$. Construct the set $\{\mathcal{I}(i)\}_{i=1}^{N_c}$ where $\mathcal{I}(i)$ is an N_f-tuple from $\{1, \ldots, M\}$ and $\mathcal{I}(i) \not\equiv \mathcal{I}(r)$ if $i \neq r$.

3. Set $\nu = 1$.

4. Find
$$\{\omega_m^*\}_{m \in \mathcal{I}(\nu)} = \arg\max \sum_{m \in \mathcal{I}(\nu)} \tilde{T}(\omega_m) x_m$$

where x_m are given by

$$\sum_{m \in \mathcal{I}(\nu)} \left|T_{l0}(e^{j\omega_m})\right|^2 x_m = \bar{K}_l \qquad l \in \mathcal{I}(\nu). \tag{7.26}$$

and the maximization is carried out over the region described by

$$x_m(\{\omega_m\}_{m \in \mathcal{I}(\nu)}) \geq 0, \qquad \forall m \in \mathcal{I}(\nu). \tag{7.27}$$

5. Let $x_m^* = x_m(\{\omega_m^*\}_{m \in \mathcal{I}(\nu)})$, $m \in \mathcal{I}(\nu)$.

If $\sum_{m \in \mathcal{I}(\nu)} |T_{l0}(e^{j\omega_m^*})|^2 x_m^* \leq \bar{K}_l \quad \forall l = 1, \ldots, M$

Stop. $\{\omega_m^*\}_{m \in \mathcal{I}(\nu)}$ is the set of optimal frequencies and $\{x_m^*\}_{m \in \mathcal{I}(\nu)}$ are the input powers on these frequencies, respectively.

Else if $\nu < N_c$ set $\nu = \nu + 1$; go to 4.

Else set $N_f = N_f + 1$; go to 2.

The search in Step 4 is over $\{\omega_m\}$ as (7.26) defines $\{x_m\}$ as an implicit function of $\{\omega_m\}$. As reported by Aoki (1971; pp.153–155) such optimization problems where the cost is described by a set of coupled equations, rather than an explicit cost function, and where the constraints are defined implicitly as in (7.27) are best tackled with the *pattern search* method.

Next, a simple example is presented to illustrate the design and the improvement on the test performance due to the input signals.

Example 7.2 Let us consider the controlled AR model

$$A(q^{-1})\, y(k) = u(k) + \epsilon(k)$$

where the white Gaussian noise $\epsilon(k)$ is of zero mean and has the variance $\sigma^2 = 1$. Let us assume that $A(q^{-1})$ has degree two, and the two hypotheses $\mathcal{H}_1$ and $\mathcal{H}_2$ are concerned with a deviation of either coefficient of $A(q^{-1})$ from those of a nominal mode $\mathcal{H}_0$; e.g.,

$$\mathcal{H}_0: \quad A(q^{-1}) = A_0(q^{-1}) = 1 - q^{-1} + 0.3q^{-2} \tag{7.28}$$

$$\mathcal{H}_1: \quad A(q^{-1}) = A_1(q^{-1}) = 1 - 0.8q^{-1} + 0.3q^{-2} \tag{7.29}$$

$$\mathcal{H}_2: \quad A(q^{-1}) = A_2(q^{-1}) = 1 - q^{-1} + 0.5q^{-2}. \tag{7.30}$$

We shall find the input signal which is optimal in the sense of (7.24)–(7.25) where we assume $\bar{K}_1 = \bar{K}_2 = 0.02$ and the input power bound is $\bar{K}_3 = K_u = 1$. The weights w_1 and w_2 are taken to be unity and $\Omega = [0, \pi]$.

To obtain the cost function in (7.24), the indices j_i^* ($i = 1, 2$) must first be determined. Since $C(q^{-1}) = 1$ regardless of which hypothesis is true, following (7.18), we have to compare

$$E\left\{\left[\frac{A_0(q^{-1})}{A_1(q^{-1})}\epsilon(k)\right]^2\right\} = \frac{1}{\pi}\int_0^\pi \left|\frac{A_0(e^{j\omega})}{A_1(e^{j\omega})}\right|^2 d\omega \tag{7.31}$$

to

$$E\left\{\left[\frac{A_2(q^{-1})}{A_1(q^{-1})}\epsilon(k)\right]^2\right\} = \frac{1}{\pi}\int_0^\pi \left|\frac{A_2(e^{j\omega})}{A_1(e^{j\omega})}\right|^2 d\omega \tag{7.32}$$

in order to determine j_i^*. Using (7.28)–(7.30) in (7.31) and (7.32) and evaluating the integrals one finds that

$$E\left\{\left[\frac{A_0(q^{-1})}{A_1(q^{-1})}\epsilon(k)\right]^2\right\} = 1.071 > 1.054 = E\left\{\left[\frac{A_2(q^{-1})}{A_1(q^{-1})}\epsilon(k)\right]^2\right\}.$$

Hence, $j_1^* = 2$ or, in other words, $E\{z_k(1,0) \mid \mathcal{H}_1\} > E\{z_k(1,2) \mid \mathcal{H}_1\}$ when $u(k) = 0$ for all k. Proceeding similarly, we also get $j_2^* = 1$. Therefore, the cost function in (7.24) is obtained as

$$\int_0^\pi \tilde{T}(\omega)\,d\xi(\omega)$$

$$= \int_0^\pi \left[|T_{21}(e^{j\omega})|^2 + |T_{12}(e^{j\omega})|^2\right] d\xi(\omega)$$

$$= \int_0^\pi \left[\left|\frac{A_2(e^{j\omega}) - A_1(e^{j\omega})}{A_1(e^{j\omega})}\right|^2 + \left|\frac{A_1(e^{j\omega}) - A_2(e^{j\omega})}{A_2(e^{j\omega})}\right|^2\right] d\xi(\omega)$$

$$= \int_0^\pi \frac{0.19 - 0.597\cos\omega + 0.662\cos^2\omega - 0.256\cos^3\omega}{1.412 - 5.99\cos\omega + 10\cos^2\omega - 7.76\cos^3\omega + 2.4\cos^4\omega}\,d\xi(\omega). \tag{7.33}$$

Next, note that
$$\left|T_{i0}(e^{j\omega})\right|^2 = \left|\frac{A_i(e^{j\omega}) - A_0(e^{j\omega})}{A_0(e^{j\omega})}\right|^2$$
and by (7.28)–(7.30), it follows that
$$\left|T_{10}(e^{j\omega})\right|^2 = \left|T_{20}(e^{j\omega})\right|^2 = \frac{0.04}{1.49 - 2.6\cos\omega + 1.2\cos^2\omega}.$$
Since $\bar{K}_1 = \bar{K}_2$, this means that two of the constraints in (7.25) are identical and the optimal input is required to satisfy
$$\int_0^\pi \left|T_{10}(e^{j\omega})\right|^2 d\xi(\omega) \le \bar{K}_1 = 0.02 \qquad (7.34)$$
and
$$\int_0^\pi d\xi(\omega) \le K_u = 1. \qquad (7.35)$$

The power spectral distribution of the optimal input can be found following the same lines as the analysis in Subsection 6.3.2. The cost function in (7.33) subject to (7.35) is maximized by
$$\xi_1(\omega) = \pi I(\omega - \omega_1)$$
where $\omega_1 = 0.889$ is the frequency maximizing $\tilde{T}(\omega)$. However, we note that
$$\int_0^\pi \left|T_{10}(e^{j\omega})\right|^2 d\xi_1(\omega) = \pi \left|T_{10}(e^{j\omega_1})\right|^2 = 0.122 \ge \bar{K}_1.$$
Therefore, $\xi_1(\omega)$ is not the optimal spectral distribution. Similarly, by analogy with (6.24)–(6.25), the integral (7.33) is maximized subject to (7.34) by
$$\xi_2(\omega) = \frac{\pi \bar{K}_1}{|T_{10}(e^{j\pi})|^2} I(\omega - \pi).$$
Nevertheless, $\xi_2(\omega)$ is not an admissible power spectral distribution either because it does not satisfy (7.35). Hence, one must search for an input which satisfies both (7.34) and (7.35) with equality.

Following the analysis in (6.27)–(6.34) and replacing $T_0(e^{j\omega})$ with $T_{10}(e^{j\omega})$ and $|T_1(e^{j\omega})|^2$ with $\tilde{T}(\omega)$, the two-dimensional search in (6.34) yields that the optimal input signal, in fact, consists of only one frequency $\omega^* = 1.753$; i.e.,
$$u(k) = \sqrt{2}\cos(1.753k + \phi), \qquad (7.36)$$

		No input	White input	Optimal input
Change	ADD	205	189	190
$\mathcal{H}_0 \to \mathcal{H}_1$	Pr{accept $\mathcal{H}_2$}	0.02	0.03	0.02
Change	ADD	242	226	215
$\mathcal{H}_0 \to \mathcal{H}_2$	Pr{accept $\mathcal{H}_1$}	0.02	0.03	0.03
No change	MTBFA	5.73×10^3	3.36×10^3	4.68×10^3

Table 7.2: Estimated detection and false alarm performances of the multihypothesis CUSUM test under offline inputs

the random phase ϕ being uniformly distributed over $[-\pi, \pi]$.

Table 7.2 gives the estimated average detection delays in detecting changes from $\mathcal{H}_0$ to $\mathcal{H}_1$ and $\mathcal{H}_2$ via a multihypothesis CUSUM test under different inputs, as well as the wrong detection detection probabilities (that is, probability of accepting $\mathcal{H}_i$ when in fact $\mathcal{H}_j$ is true). The estimated mean times between the false alarms are also tabulated. The results are based on 1000 runs with thresholds $\bar{\beta}_{ij} = 4$ ($i = 1, 2$, $j = 0, 1, 2$, $i \neq j$). In estimating the detection delay the system is run under $\mathcal{H}_0$ until $k = 200$, when the change is introduced. The power of the auxiliary white Gaussian signal is chosen equal to that of the optimal input in (7.36), i.e., unity.

It is seen from Table 7.2 that the optimal input in (7.36) can accelerate the detection of either mode at least as much as white noise can. Both of them decrease the mean time between false alarms while improving the detection time. However, the optimal input yields larger false alarm times than the other one. Neither input seems to have a significant effect on the probabilities of wrong detection.

7.3.3 Online inputs

Before concluding this chapter we shall briefly outline the generalization of suboptimal online inputs which have been discussed in Chapter 6 for the two-hypotheses case.

By extending the line of thought in Subsection 6.4.2 in view of (7.21) and (7.22), the design objective in the multihypothesis case can be taken so as to achieve

(7.21) using the feedback law $u(k) = F(q^{-1})y(k)$ without affecting the quantities $E\{z_k(i,0) \mid \mathcal{H}_0\}$ ($i = 1, \ldots, M-1$) and $E\{y^2(k) \mid \mathcal{H}_j\}$ ($j = 0, \ldots, M-1$).

The average increments of various cumulative sums under output feedback can be obtained from (7.14) with manipulations similar to (6.44)–(6.47) as

$$E\{z_k(i,0) \mid \mathcal{H}_0\} = -\frac{1}{2\sigma^2}E\left\{\left[\frac{\bar{A}_{i0}(q^{-1}) - q^{-d}\bar{B}_{i0}(q^{-1})\,F(q^{-1})}{(A_0(q^{-1}) - q^{-d}B_0(q^{-1})\,F(q^{-1}))\,C_i(q^{-1})}\epsilon(k)\right]^2\right\}$$

and

$$E\{z_k(i,j) \mid \mathcal{H}_i\} = \frac{1}{2\sigma^2}E\left\{\left[\frac{\bar{A}_{ij}(q^{-1}) - q^{-d}\bar{B}_{ij}(q^{-1})\,F(q^{-1})}{(A_i(q^{-1}) - q^{-d}B_i(q^{-1})\,F(q^{-1}))\,C_j(q^{-1})}\epsilon(k)\right]^2\right\}.$$

where

$$\bar{A}_{ij}(q^{-1}) = A_i(q^{-1})\,C_j(q^{-1}) - A_j(q^{-1})\,C_i(q^{-1})$$

and

$$\bar{B}_{ij}(q^{-1}) = B_i(q^{-1})\,C_j(q^{-1}) - B_j(q^{-1})\,C_i(q^{-1})$$

for $i = 1, \ldots, M-1$ and $j = 0, \ldots, M-1$ ($i \neq j$).

By linearizing these quantities around $F(q^{-1}) = 0$ and considering the approximations for the output variance in (6.52), one can proceed following the same lines as Subsection 6.4.2. Eventually, the optimization problem which reflects the above-mentioned objectives up to first order approximations will be similar to (6.53) except that it will involve $2M - 1$ orthogonality constraints. Namely,

$$\begin{array}{ll}\underset{\mathbf{f}}{\text{maximize}} & \mathbf{f}^T\mathbf{s} \\ \text{subject to} & \mathbf{f}^T\mathbf{r}_1 = \cdots \mathbf{f}^T\mathbf{r}_{2M-1} = 0 \quad \text{and} \quad \mathbf{f}^T\mathbf{f} \leq K_f.\end{array}$$

Similar to (6.54), the $\mathbf{s}$ vector contains the factors of the coefficients of $F(q^{-1})$ in linear approximation of the cost function in (7.21). The vectors $\mathbf{r}_l$ ($l = 1, \ldots, 2M - 1$) are analogously related to the approximations of $E\{z_k(i,0) \mid \mathcal{H}_0\}$, ($i = 1, \ldots, M - 1$) and $E\{y^2(k) \mid \mathcal{H}_j\}$ ($j = 0, \ldots, M - 1$). The suboptimal coefficients of the feedback polynomial can then be found by (6.58) where $\mathbf{R}$ is to be taken as $\mathbf{R} = [\mathbf{r}_1 \cdots \mathbf{r}_{2M-1}]$. Nevertheless, one should note that $F(q^{-1})$ is required to be of degree $2M$ in the multihypothesis case if the false alarms and output variances related to all the hypotheses are to be taken into account.

7.4 Conclusions

In this chapter, the change detection and input design techniques discussed in Chapters 5 and 6 are extended to the case where there are more than two hypotheses describing the operation mode of a dynamic process.

The CUSUM test is generalized to the multihypothesis case by considering its equivalence to repeating SPRT's until $\mathcal{H}_0$ is rejected. This new test is contrasted to another extension of the CUSUM test proposed by Zhang (1989). Despite the lack of information about the average run length of the multihypothesis CUSUM test, inputs improving the test performance can still be designed by considering the effect of offline inputs on the average increments of various cumulative sums used by the test.

References

Albert, G.E. (1947). A note on the fundamental identity of sequential analysis, *Annals of Mathematical Statistics*, **18**, 593–596 (Correction in **19**, 426–427).

Aoki, M. (1971). *Introduction to Optimization Techniques: Fundamentals and Applications of Nonlinear Programming*, Macmillan, New York.

Armitage, P. (1950). Sequential analysis with more than two alternative hypotheses, and its relation to discriminant function analysis, *Jour. Royal Statistical Society — B*, **12**, 137–144.

Baskiotis, C., J. Raymond and A. Rault (1979). Parameter identification and discriminant analysis for jet engine mechanical state diagnosis, *Proc. 18th IEEE Conference on Decision and Control*, Fort Lauderdale.

Basseville, M. (1986). On-line detection of jumps in the mean, in *Detection of Abrupt Changes in Signals and Dynamical Systems*, ed. M. Basseville and A. Benveniste, *Lecture Notes in Control and Information Sciences*, **77**, Springer, Berlin.

Basseville, M. (1988). Detecting changes in signals and systems—a survey, *Automatica*, **24**, 309–326.

Basseville, M. and A. Benveniste (1983). Sequential detection of abrupt changes in spectral characteristics of digital signals, *IEEE Trans. on Information Theory*, **IT-29**, 709–724.

Basseville, M. and A. Benveniste (1986). *Detection of Abrupt Changes in Signals and Dynamical Systems*, *Lecture Notes in Control and Information Sciences*, **77**, Springer, Berlin.

Bellman, R. (1957). On a generalization of the fundamental identity of Wald, *Proc. Cambridge Philosophical Society*, **53**, 257–259.

Benveniste, A. (1986). Advanced methods of change detection: An overview, in *Detection of Abrupt Changes in Signals and Dynamical Systems*, ed. M. Basseville and A. Benveniste, *Lecture Notes in Control and Information Sciences*, **77**, Springer, Berlin.

Bertsekas, D.P. (1987). *Dynamic Programming: Deterministic and Stochastic Models*, Prentice-Hall, New Jersey.

Blackwell, D. and M.A. Girshick (1954). *Theory of Games and Statistical Decisions*, John Wiley & Sons, New York.

Box, G.E.P. and G.M. Jenkins (1976). *Time Series Analysis: Forecasting and Control*, Holden-Day, San Francisco.

Brumback, B.D. and M.D. Srinath (1987). A chi-square test for fault detection in Kalman filters, *IEEE Trans. on Automatic Control*, **AC-32**, 552–554.

Chen, M.J. and J.P. Norton (1987). Estimation technique for tracking rapid parameter changes. *International Jour. of Control*, **45**, 1387–1398.

Chien, T.T. and M.B. Adams (1976). A sequential failure detection technique and its application, *IEEE Trans. on Automatic Control*, **AC-21**, 750–757.

Chow, E.W. and A.S. Willsky (1984a). Analytical redundancy and the design of robust failure detection systems, *IEEE Trans. on Automatic Control*, **AC-29**, 603–614.

Chow, E.W. and A.S. Willsky (1984b). Bayesian design of decision rules for failure detection, *IEEE Trans. on Aerospace and Electronic Systems*, **AES-20**, 761–773.

Clark, R.N. (1978). A simplified instrument failure detection scheme, *IEEE Trans. on Aerospace and Electronic Systems*, **AES-14**, 558–563.

Clark, R.N. and W. Setzer (1980). Sensor fault detection in a system with random disturbances, *IEEE Trans. on Aerospace and Electronic Systems*, **AES-16**,

468–473.

Deckert, J.C., M.N. Desai, J.J. Deyst and A.S. Willsky (1977). F-8 DFBW sensor failure identification using analytical redundancy, *IEEE Trans. on Automatic Control*, **22**, 795–803.

DeGroot, M. (1970). *Optimal Statistical Decisions*, McGraw-Hill, New York.

Doob, J.L. (1953). *Stochastic Processes*, John Wiley & Sons, New York.

Dowdle, J.R., A.S. Willsky and S.W. Gully (1983). Nonlinear generalized likelihood ratio algorithms for maneuver detection and estimation, *Proc. American Control Conference*, Arlington.

Eisenberg, B. and B.K. Ghosh (1976). Properties of generalized sequential probability ratio tests, *The Annals of Statistics*, **4**, 237–251.

Eisenberg, B. and B.K. Ghosh (1979). The likelihood ratio and its applications in sequential analysis, *Jour. Multivariate Analysis*, **9**, 116–129.

Ferguson, T.S. (1967). *Mathematical Statistics: A Decision Theoretic Approach*, Academic Press, New York.

Fleisher, S. and E. Shwedyk (1980). A sequential multiple hypothesis test for the unknown parameters of a Gaussian distribution, *IEEE Trans. on Information Theory*, **IT-26**, 255–259.

Frank, P.M. and L. Keller (1980). Sensitivity discriminating observer design for instrument failure detection, *IEEE Trans. on Aerospace and Electronic Systems*, **AES-16**, 460–467.

Frank, P.M., B. Köppen and J. Wünnenberg (1991). General solution of the robustness problem in linear fault detection filters, *Proc. 1st European Control Conference*, Grenoble.

Geiger, G. (1984). Fault identification of a motor pump system using parameter estimation and pattern classification, *Proc. 9th IFAC World Congress*, Budapest.

Gertler, J.J. (1988). Survey of model based failure detection and isolation in complex plants, *IEEE Control Systems Magazine*, **M-CS-8**, 3–11.

References

Ghosh, B.K. (1970). *Sequential Tests of Statistical Hypotheses*, Addison Wesley, Reading.

Goodwin, G.C. and R.L. Payne (1977). *Dynamic System Identification: Experiment Design and Data Analysis*, Academic Press, New York.

Grenander, U. and G. Szegö (1958). *Toeplitz Forms and Their Applications*, Univ. California Press, Berkeley.

Gustafson, D.E., A.S. Willsky, J.Y. Wang, M.C. Lancaster and J.H. Triebwasser (1978). ECG/VCG rhythm diagnosis using statistical signal analysis I: Identification of persistent rhythms. *IEEE Trans. on Biomedical Engineering*, **BME-25**, 344–353.

Hallenbeck, D.J. and T.H. Macgregor (1984). *Linear Problems and Convexity Techniques in Geometric Function Theory*, Pitman, London.

Hannan, E.J. (1970). *Multiple Time Series*, John Wiley & Sons, New York.

Himmelblau, D.M. (1978). *Fault Detection and Diagnosis in Chemical and Petrochemical Systems*, Elsevier, Amsterdam.

Homssi, L. and A. Despujols (1991). Fault diagnosis based on modelling and parameter identification, *Proc. 1st European Control Conference*, Grenoble.

Isermann, R. (1984). Process fault detection based on modelling and estimation methods—a survey, *Automatica*, **20**, 387–404.

Johnson, N.L. (1961). Sequential analysis: A survey, *Jour. Royal Statistical Society — Part A*, **124**, 372–411.

Kalaba, R. and K. Spingarn (1982). *Control, Identification and Input Optimization*, Plenum, New York.

Kerestecioğlu, F. and M.B. Zarrop (1989). Bayesian approach to optimal input design for failure detection and diagnosis, *Proc. IFAC Symposium on Adaptive Systems in Control and Signal Processing*, Glasgow.

Kerestecioğlu, F. and M.B. Zarrop (1990). Sequential analysis of stationary autoregressive processes, Control Systems Centre Report No. 725, UMIST, Manchester.

Kerestecioğlu, F. and M.B. Zarrop (1991). Optimal input design for change detection in dynamical systems, *Proc. 1st European Control Conference*, Grenoble.

Kullback, S. (1959). *Information Theory and Statistics*, John Wiley & Sons, New York.

Lorden, G. (1971). Procedures for reacting to a change in distribution, *Annals of Mathematical Statistics*, **42**, 1897–1908.

Lou, X., A.S. Willsky and G.C. Verghese (1986). Optimally robust redundancy relations for failure detection in uncertain systems. *Automatica*, **22**, 333–344.

Mehra, R.K. (1974). Optimal inputs for linear system identification, *IEEE Trans. on Automatic Control*, **19**, 192–200.

Mentz, R.P. (1976). On the inverse of some covariance matrices of Toeplitz type, *SIAM Jour. Applied Mathematics*, **31**, 426–437.

Miller, H.D. (1962). Absorption probabilities for some of random variables defined on a finite Markov chain, *Proc. Cambridge Philosophical Society*, **58**, 286–298.

Minorovskii, L.A. (1981). Functional diagnosis of dynamic systems (Survey), *Automation and Remote Control*, **41**, 1122–1143.

Moustakides, G.V. (1986). Optimal stopping times for detecting changes in distributions, *Annals of Mathematical Statistics*, **14**, 1379–1387.

Newbold, P.M. and Y.C. Ho (1968). Detection of changes in the characteristics of a Gauss-Markov process, *IEEE Trans. on Aerospace and Electronic Systems*, **AES-4**, 707–718.

Nikiforov, I.V. (1979). Cumulative sums for detection of changes in random process characteristics, *Automation and Remote Control*, **40**, 192–200.

Nikiforov, I.V. (1986). Sequential detection of changes in stochastic systems, in *Detection of Abrupt Changes in Signals and Dynamical Systems*, ed. M. Basseville and A. Benveniste, *Lecture Notes in Control and Information Sciences*, **77**, Springer, Berlin.

Page, E.S. (1954). Continuous inspection schemes, *Biometrika*, **41**, 100–115.

Patton, R.J. and J. Chen (1991). A parity space approach to robust fault detection using eigenstructure assignment, *Proc. 1st European Control Conference*, Grenoble.

Patton, R., P. Frank and R. Clark (1989). *Fault Diagnosis in Dynamic Systems: Theory and Applications*, Prentice Hall, Hemel Hempstead.

Patton, R.J. and S.M. Kangethe (1989). Robust fault diagnosis using eigenstructure assignment of observers, in *Fault Diagnosis in Dynamic Systems: Theory and Applications*, ed. Patton, R., P. Frank and R. Clark, Prentice Hall, Hemel Hempstead.

Pau, L.F. (1981). *Failure Diagnosis and Performance Monitoring*, Marcel Dekker, New York.

Pesaran, M.H. and L.J. Slater (1980). *Dynamic Regression: Theory and Algorithms*, Ellis Horwood, Chichester.

Phatarfod, R.M. (1965). Sequential analysis of dependent observations. I, *Biometrika*, **52**, 157–165.

Phatarfod, R.M. (1971). Sequential tests for normal Markov sequence, *Jour. Australian Mathematical Society*, **2**, 433–440.

Prudnikov, A.P., Y.A. Brychov and O.I. Marichev (1986). *Integrals and Series*, Gordon and Breach Science, New York.

Reynolds, M.R. (1975). Approximations to the average run length in cumulative sum control charts, *Techometrics*, **17**, 65–71.

Roebuck, P.A. and S. Barnett (1978). A survey of Toeplitz and related matrices, *International Jour. Systems Science*, **9**, 921–934.

Shapiro, J.F. (1979). *Mathematical Programming: Structures and Algorithms*, John Wiley & Sons, New York.

Shiryaev, A.N. (1978). *Optimal Stopping Rules*, Springer, New York.

Siegmund, D. (1985). *Sequential Analysis: Tests and Confidence Intervals*, Springer, New York.

Sobel, M. and A. Wald (1949). A sequential decision procedure for choosing one of three hypotheses concerning the unknown mean of a normal distribution, *Annals of Mathematical Statistics*, **20**, 502–522.

Therrien, C.W., T.F. Quatieri and D.E. Dudgeon (1986). Statistical model-based algorithms for image analysis, *Proc. IEEE*, **74**, 532–551.

Tweedie, M.C.K. (1960). Generalization of Wald's Fundamental Identity of sequential analysis to Markov chains, *Proc. Cambridge Philosophical Society*, **56**, 205–214.

Uosaki, K., I. Tanaka and H. Sujiyama (1984). Optimal input design for autoregressive model discrimination with constrained output variance, *IEEE Trans. on Automatic Control*, **AC-29**, 348–350.

Uosaki, K. and T. Hatanaka (1987). Optimal input design for autoregressive model discrimination based on the Kullback-Leibler discrimination information, *Proc. IFAC 10th Triennial World Congress*, Munich.

Upadhyaya, B.R. and H.W. Sorenson (1977). Synthesis of linear stochastic signals in identification problems, *Automatica*, **13**, 615–622.

Wald, A. (1947). *Sequential analysis*, John Wiley & Sons, New York.

Wald, A. and J. Wolfowitz (1948). Optimum character of the sequential probability ratio test, *Annals of Mathematical Statistics*, **19**, 326–339.

Walsh, G.R. (1971). *An Introduction to Linear Programming*, Holt, Reinhardt and Winston, London.

Wellstead, P.E. and M.B. Zarrop (1991). *Self-Tuning Systems: Control and Signal Processing*, John Wiley & Sons, Chichester.

Wetherill, G.B. and K.D. Glazebrook (1986). *Sequential Methods in Statistics*, Chapman and Hall, London.

Willsky, A.S. (1976). A survey of design methods for failure detection in dynamic systems. *Automatica*, **12**, 601–611.

Willsky, A.S. (1986). Detection of abrupt changes in dynamic systems, in *Detection* of *Abrupt Changes in Signals and Dynamical Systems*, ed. M. Basseville and A. Benveniste, *Lecture Notes in Control and Information Sciences*, **77**, Springer, Berlin.

Willsky, A.S., E.Y. Chow, S.B. Gershwin, C.S.Greene, P.K. Houpt and A.L. Kurkjian (1980). Dynamic model-based techniques for the detection of incidents on freeways, *IEEE Trans. on Automatic Control*, **AC-25**, 347–360.

Willsky, A.S. and H.L. Jones (1976). A generalized likelihood ratio approach to the detection and estimation of jumps in linear systems, *IEEE Trans. on Automatic Control*, **AC-21**, 108–112.

Yuan, Z.D. and L. Ljung (1984). Black-box identification of multivariable transfer functions—asymptotic properties and optimal input design, *International Jour. of Control*, **40**, 233–256.

Zarrop, M.B. (1979). *Optimal Experiment Design for Dynamic System Identification, Lecture Notes in Control and Information Sciences*, **21**, Springer, Berlin.

Zhang, X.J. (1989). *Auxiliary Signal Design in Fault Detection and Diagnosis, Lecture Notes in Control and Information Sciences*, **134**, Springer, Berlin.

Zhang, X.J. and M.B. Zarrop (1988). Auxiliary signals for improving on-line fault detection, *Proc. Control 88*, Oxford.

Index

Adaptive control, 2
Analytical redundancy, 3
ARL, see average run length
ASN, see average sample number
Average detection delay, 86, 98
Average run length, 86
    approximate formulae, 87, 88
    in testing first order AR parameter, 89
Average sample number, 43, 44–45, 49
    approximate formulae, 45
    autoregressive case, 67–69
        approximate formulae, 69, 71
    in testing first order AR parameter, 74
Autoregressive model, 8, 52, 93
Autoregressive moving average process, 28, 76, 89

Bayesian decision theory, 9, 13
Bayes risk, 20
    minimum conditional, 24

Controlled autoregressive moving average model, 11, 83, 130

Cumulative sum algorithm, 6
Cumulative sum test, 80–85
    multihypothesis, 124–130
    optimality property of, 6, 86
    performance of, 85–88

Decision making, 2, 3, 4–7
Decision rule(s), 14, 18, 39
    Bayesian, 20, 25
    dynamic programming solution for, 20–27
    including input law, 15–20
    nonrandomized, 19
    optimal, 14
    pure, 19
    sequential, 14, 18, 19, 20, 34
Doubly symmetric matrix, 55
Dynamic programming, 9, 20

Error probabilities, 6, 28, 42, 123
Estimate for the change time, 83
Extreme point, 99, 118, 119, 120

False alarm probability, 41, 85, 94
Fault detection and diagnosis, 2
Fundamental identity, 10, 43–44

Index 150

analogue of, in the autoregressive case, 10, 66–67
extensions to, 53
Gauss-Markov process, 53, 71, 76
Generalized likelihood ratio test, 7
Hardware redundancy, 2
Hypothesis testing, 39, 93
    M-ary, 13, 15
Image processing, 2
Input design, 8-9
    for change detection, 8, 92–120
    for hypothesis testing, 93
    for model discrimination, 8
    for system identification, 8
    multihypothesis, 131–139
Input law, 19
    biassed, 21
Instrument fault detection, 2
Kalman filter, 4, 5
Kullback divergence, 84
Lagrange multipliers, 112
Likelihood ratio, 6, 10, 33
    log, 28, 69, 122
        conditional, 53
            asymptotic behaviour of, 10, 54–62
        increments of, 41, 52, 131

moment generating function of, 10, 43, 95
initializing, 42
moment generating function of, 54, 55, 70
Linear programming, 99
Luenberger observer, 3
Mean time between false alarms, 11, 86, 94, 98, 114, 133
Missed alarm probability, 41, 85
Model discrimination, 8, 80
Moment generating function,
    of increments of log likelihood ratio, 10, 43, 95
    of log likelihood ratio, 54, 55, 70
Multiple model approach, 5
OC, see operating characteristics
Offline inputs, 8, 11, 96–105
    multihypothesis case, 133–138
    optimal, 99, 134
        design procedure to obtain, 134
Online inputs, 11, 105–112
    multihypothesis case, 138–139
    suboptimal solution for, 108–112, 139
Operating characteristics, 43, 44
    approximate formula, 44
    autoregressive case, 67–69

Index 151

approximate formula, 69, 71
   in testing first order AR parameter, 75
Optimal input design problem, 94
   offline case, 98, 133
   online case, 108
Output feedback, 11, 92, 105

Parameter estimation, 4
Pattern search, 136
Prediction error, 84
   under offline input, 97, 130
   under online input, 107
Probability ratio, 5

Quasi-SPRT, 33

Residuals, 16
   generation of, 2–4
Robustness, 3
Risk function, 14, 20

Sequential analysis, 40
   of autoregressive processes, 51–79
Sequential decision theory, 5, 9, 39
Sequential decision problem, 7, 13, 14, 19
   elements of, 15-17
   truncated, 23, 25
Sequential probability ratio test, 5, 9, 10, 11, 28, 40–50

   definition of, 28, 41
   for autoregressive models, 52
   generalized, 33
   multihypothesis, 11, 121–124
      thresholds of, 123–124
   optimality property of, 6, 28, 40
   termination property of, 41, 64
   thresholds of, 42
Set of states of nature, 15
Set of terminal decisions, 16
Signal processing, 2
Software redundancy, 3
Spectral distribution function, 96
   onesided, 96
Spectrum, 96
SPRT, see sequential probability ratio test
Stopping rule, 14, 18
   augmented, 19, 20, 21, 23–25
   biased, 22
   optimal, 24, 37
Suboptimal solution for online inputs, 108–112, 139
Supporting hyperplane, 99
System identification, 8

Terminal decision loss, 16, 22
Terminal decision rule, 14, 19
   optimal, 20–23

Index

Toeplitz:
    determinants, asymptotic behaviour of, 56
    matrix, 55, 56, 57, 61
        associated, 55, 57, 77
        nearly, 55, 56, 57, 59, 77

Type I error, 41

Type II error, 41

Unit step function, 100

Wald's identity, 44, 53

Zhang's procedure, 125, 126, 129